Online Searching
on DIALOG®

BEILSTEIN

Reference Manual

Stephen R. Heller
US Agricultural Research Service

George W. A. Milne
US National Institutes of Health

Springer

Dr. Stephen R. Heller
2413 Lillian Drive
Silver Spring, MD 20902
USA

William A. Milne
9520 Linden Ave.
Bethesda, MD 20014
USA

ISBN 978-3-662-09080-0   ISBN 978-3-662-09078-7 (eBook)
DOI 10.1007/978-3-662-09078-7

50/3020-543210 – Printed on acid-free paper

Preface

This manual is meant to be one of the first steps in the renaissance of the *Beilstein Handbook of Organic Chemistry* and its computer-readable counterparts, the Beilstein Databases of factual and structural data. The enormous work of the staff of the Beilstein Institute has produced, for over 100 years, a very valuable and unique scientific resource. We are pleased to be able to be involved in making this large volume of evaluated scientific data more readily available to the worldwide chemical community.

We would like to thank the many staff members of the Beilstein Institute for their help in providing us with the necessary information, facts, and corrections to this manual. In particular we would like to thank Clemens Jochum, Reiner Luckenbach, Sandy Lawson, Laszlo Domokos, Martin Hicks, Steve Welford, and especially Christiane Schaum and Gabriele Ilchmann of the Beilstein Institute. We are also indebted to many teachers of organic chemistry and colleagues in the field of computers and chemical information, including Fausto Ramirez, Ed Kosower, Chuck Hammer, Richard Feldmann, and Chezi Wolman.

Stephen R. Heller
Silver Spring, Maryland

George W. A. Milne
Bethesda, Maryland
June 1990

Contents

CONTENTS

CONTENTS

CONTENTS

I. Introduction to Beilstein Online

The standard reference work known today as *Beilsteins Handbuch der Organischen Chemie* is a descendant of the original *Handbuch*, whose first edition was created by F. K. Beilstein in St. Petersburg in 1881.

Beilstein, born in St. Petersburg of German parents in 1838, was educated at the Universities of Heidelberg, Munich, and Göttingen and assumed a professorship at the Imperial Technical Institute in St. Petersburg in 1866. The first edition of *Beilsteins Handbuch der Organischen Chemie* was published in 1881–1882 and consisted of two volumes with a total of 2200 pages in which about 15,000 organic compounds were described. Beilstein published a second edition (three volumes, 4080 pages) between 1885 and 1889, and a third edition (eight volumes, 11,000 pages) between 1892 and 1906, the year of his death.

It was clear by this time that the size and scope of the *Handbook* were such that it could no longer be managed by an individual, and accordingly, the German Chemical Society undertook this responsibility following Beilstein's death. Publication of the current edition of *Beilstein*, the fourth edition, began in 1918, under the editorship of P. Jacobson and B. Prager. In 1933, F. Richter was named as editor, and he was followed in 1961 by H.-G. Boit. The current editor, R. Luckenbach, succeeded Professor Boit in 1978.

The fourth edition of *Beilstein* is the basis of Beilstein Online. It consists of a main work (*Hauptwerk*) and five supplementary series (*Ergänzungswerke*). The combined supplement EIII/EIV is not regarded as a separate supplement. Each of these supplementary series covers different time periods as shown in Table I. The basic work and the first four supplements were published in German; the fifth supplement is in English.

The main work and each of the supplementary series consist of 27 volumes, each of which may be one or more physical books. Which compounds appear in which volume is determined by the compound's chemical structure, as is described below. A given compound will appear in the same volume in each series; thus thiophene is found in volume 17 of the main work and volume 17 of each of the supplementary series, because volume 17 is devoted to heterocyclics containing one chalcogen (Group VI: O, S, Se, or Te) heteroatom. The main consequence of this organization is that the *Beilstein Handbook*, rather than being a linear chronological record, is really a series of snapshots, taken at 10 or 20 year intervals, of the entire organic chemical world.

A central feature of the *Handbook*, therefore, is the way in which it is organized. A specific structure will be found at essentially the same place in any of the supplementary series. Determining that location from the structure always required a knowledge of the Beilstein systematic rules for filing. These rules are well described in a brochure published by the Beilstein Institute and entitled *How to Use Beilstein*. This and other brochures on the *Beilstein Handbook* are available from either the Beilstein Institute or Springer-Verlag Publishers (see Appendix A for the nearest source to obtain this information). They are not covered here because Beilstein Online, with its searching capability, will retrieve *Beilstein* locations for you. When a chemical is retrieved from Beilstein Online, the *Beilstein* citations—typically one per series—are provided. These appear in the form 4-17-00-00093, which indicates page 93 of volume 17 of the fourth supplementary series.

Most of the factual data in the *Handbook* are also accessible online, but recourse to the printed *Handbook* may be advisable, or even necessary, for compounds associated with large numbers of data. For example, the complete printout of all data in Beilstein Online for the chemical compound pyridine requires some 244 pages.

TABLE I. Beilstein Handbook, fourth edition

Basic Series	1830–1909	1
Supplement I	1910–1919	E I
Supplement II	1920–1929	E II
Supplement III	1930–1949	E III
Supplement III, IV	1930–1959	E III/IV
Supplement IV	1950–1959	E IV
Supplement V	1960–1979	E V

The heart of the *Beilstein Handbook* is the factual information associated with each compound. Each series contains new information about a compound, and in the online database, data from the main work and all the supplementary series are combined to form a record which encompasses the available knowledge about the structure in question. For each compound in the Beilstein Database data are provided, as available, on:

- Chemical Structure

- Chemical Name

- Beilstein Registry Number

- Beilstein Citation

- Molecular Formula

- Molecular Weight

- Lawson Number

- Chemical Reactions and Preparation

- Isolation from Natural Products

- Structure and Energy Parameters

- Physical State

- Mechanical Properties

- Calorific Data

- Optical Data

- Magnetic and Electrical Properties

- Electrochemical Behavior

- Multi-Component System Data

- Spectral Data

- Transport Phenomena

- Physiological Behavior and Applications

The total number of types of data that may be available for a compound is in excess of 400. The diagram in section 2B15 presents a graphical overview of the entire Beilstein Database. Each of these fields may be retrieved and displayed, and almost all of them may be used in searching. It should be noted that very few compounds have many fields associated with them. To get an idea of how many entries have a particular type of data, refer to the Data Present (DP=) information in Appendix D.

In addition to all the factual data, which are extensively reviewed and cross-checked before they are added to the database, a record is kept of the source of the data. A literature citation for every measurement of a datum which appears in this database is provided. It is possible also to search through these literature citations, on the basis, for example, of authors' surnames.

II. Introduction to DIALOG Search Software

Introduction and Purpose
General System Information

2A. INTRODUCTION AND PURPOSE

This chapter is designed to provide you with a brief introduction to the DIALOG search software system capabilities and features. It is not meant to replace the extensive and detailed *Searching DIALOG: The Complete Guide*, available from Dialog Information Services.

The DIALOG search software is functionally similar to that of STN-International Messenger and ORBIT, two other systems which are familiar to many users. Some of the commands are the same, while others are slightly or considerably different. In this chapter you will be only given the information on how to use the DIALOG software, not how to conduct searches of the Beilstein database. Chapter 3 of this manual is designed to show you how to search for factual data in the Beilstein Database using the DIALOG software.

2B. General System Information

The procedures to obtain an account and to log in to the DIALOG system are given in Appendix A. When you log in to DIALOG, you are normally, by default, put into either file 1 (ERIC) or file 204 (ONTAP). You may conduct searches in this file, or you can begin any other file from the 300–400 offered by DIALOG. The Beilstein Database is in file 390 in the DIALOG system, and you can enter this file with the command BEGIN 390, or B 390. If you want to be automatically connected to the Beilstein Database every time you log in, you may request this by contacting DIALOG Customer Administration in writing or on DIALMAIL (To: CUSTADMIN). DIALMAIL is DIALOG's electronic mail system. It is described in detail in Appendix D of *Searching DIALOG: The Complete Guide*. The general and administrative chores you can do while in the default file (or any other file) include:

- Read general HELP file messages
- Read general NEWS files
- Use DIALMAIL to send a message to DIALOG

In the default file you can also LOGOFF and leave DIALOG, and of course you can switch to the Beilstein File by typing BEGIN 390.

2B1. DIALOG Prompts

When you log in to DIALOG, the first thing you see after the login and news messages is the prompt sign (?), which lets you know the system is waiting for you to do something.

2B2. DIALOG Command Pattern

In the DIALOG system there is a general form which you use to enter all commands. There are five parts to this command pattern, which looks like this:

$$\frac{? \quad 2 \qquad 4 \quad (CR)}{1 \quad 2 \quad 3 \quad 4 \qquad 5}$$

where

- 1   is the system prompt sign;
- 2   is the actual command;
- 3   is a blank space which is needed to separate items 2 and 4;
- 4   is the information you are associating with the command (item 2);
- 5   is the carriage return or "enter" key you must strike to assure the command string is transmitted to the computer.

A typical command, using all these parts, is:

```
? s mp=123 (CR)
```

The command s, short for select, indicates that you wish to carry out a search. The mp=123 is the information for which you want to search, and the command string is ended, as all commands must be, with a carriage return.

2B3. Set Numbers (S#'s)

In order to keep track of each command you use in the system, the system creates a list of numbers associated with each question you ask, structure you create, or set of answers you obtain after a search is performed. These lists are specified by means of S-numbers, or S#'s. These S#'s are essentially lists of retrieved records, and they can be used during the search session in any combination with other S#'s or commands as you wish. The DIALOG search software provides for internal space to hold up to 200 S#'s at one time. After you reach 199 S#'s, the system will ask you if you wish to delete some or all of the 199 S#'s you have created. After you create S199, you cannot continue searching unless you delete at least one S#.

2B4. System Limits

Table II lists the current system limitations for searching of both the factual data and structure files. These limits are arbitrary and may be changed at any time; they are designed to assure reasonable response time for all users. When you reach any limit there are a number of things you may do. You may divide your search into parts, either by refining the search or by using the range command with Beilstein Registry Numbers for example, or with a range of carbon atoms, a range of molecular weights, and so on.

TABLE II. DIALOG system limits

Maximum number of occurrences of all terms in a single search	5,000,000
Maximum number of answers in an EXPAND command	50
Maximum number of characters in a single command	240
Maximum number of characters in a word or phrase (including prefix code)	49
Maximum number of prefixes allowed with a given search term	7
Maximum number of characters per series of suffix codes	40
Maximum number of logical operations in a single command	49
Maximum number of S, E, or R numbers entered using OR or :	50
Maximum number of S, E, or R numbers entered using commas	7
Maximum number of terms retrieved with colon or truncation	2,000
Maximum number of index entries retrieved with truncation	20,000
Maximum number of characters in type or display commands	240
Maximum number of item ranges in type or display commands	39
Maximum number of print commands per day	999
Maximum number of characters per print command	240
Maximum number of records per print command	5,000
Number of records printed if no item range is provided	50
Maximum number of parameters per print command	85
Maximum number of item ranges per print command	45
Maximum number of characters in the title using print title	80
Maximum hold (in hours) on print commands	2
Maximum number of transaction numbers per print cancel command	10
Maximum S#'s allowed in concurrent use	200
Maximum number of characters in a SELECT statement	240
Maximum length of a single term	49
Maximum number of Beilstein records which can be printed per S#	5,000
Maximum time, in minutes, to re-log in (and retain results files) after being disconnected from the system	10
Maximum idle time, in minutes, online before automatic logoff	10

2B5. Help Messages

The DIALOG search software has many help and error messages built into the system. The HELP or EXPLAIN command can be used to view help messages that describe DIALOG commands, system features, file-specific information, and training information. For example, to view a description of the BEGIN command, enter HELP BEGIN or EXPLAIN BEGIN. The HELP/EXPLAIN command can be abbreviated with a question mark, e.g., ?BEGIN.

The valid HELP/EXPLAIN options are listed below.

- DIALOG commands (e.g., HELP BEGIN):

BEGIN	FILE	PAGE	SELECT
COST	KEEP	RD	SET
DISPLAY	ID	RELEASE	SFILES
DS	MAP	PRINT	SHOW
EDIT	LOGOFF	RECALL	SORT
EXECUTE	ORDER	REPORT	TYPE
EXPAND	LIMITALL	SAVESDI	

- Special features (e.g., HELP HILIGHT):

ALERT	LOGICAL	SUBACCT
DUP	ONESEARC	TAG
HILIGHT	POST	THESAURI
KWIC	PROFILE	TRUNCATE
LABELS	PROXIMIT	XTAB

- DIALOG news and system status (e.g., HELP DIALINDX):

DIALINDX	FILESUM	NEWS	SCHEDULE	UPDATE
FILES	FREE	ONTAP	SUPPLRS	
FILESAZ	INSTRUCT	RATES	TOLLFREE	

- For telecommunication access information (e.g., HELP ACCESS):

ACCESS	DIALNET	SABD	TELEX	WATS
DARDO	IDAS	TELENET	TRANSPAC	
DATAPAC	NORPAC	TELEPAC	TWX	
DATEX	PSS	TELEPAKS	TYMNET	

- For file information (e.g., HELP FIELD114):

        FIELD*n* FILE*n* FMT*n* LIMIT*n* RATES*n* RPT*n* SORT*n* MAP*n*

    For information about DIALOG training seminars, enter HELP TRAIN.
    For information about training provided by Database Suppliers, enter:

HELP ANZNEWS	(Australia/New Zealand)
HELP CANNEWS	(Canada)
HELP EURNEWS	(Europe)
HELP KINONEWS	(Kinokuniya Japan)
HELP MEXNEWS	(Mexico)
HELP MMCNEWS	(Masis Japan)
HELP USNEWS	(United States)

2B6. Online News

DIALOG publishes a monthly newsletter called *CHRONOLOG*, and the text of this
newsletter is online as File 410. To read news in *CHRONOLOG* about the Beilstein
Database, begin 410 and then select 390 or select beilstein. This will produce a results file which you can browse through (using format 5):

```
?b 410
        05dec89 14:46:54 User 054911 Session B64.2
            $1.52    0.101 Hrs File415
               $0.00  1 Type(s) in Format  5
            $0.00  1 Types
    $1.52  Estimated cost File415
    $1.11  Telenet
    $2.63  Estimated cost this search
    $2.84  Estimated total session cost   0.105 Hrs.

File 410:CHRONOLOG NEWSLETTER -JAN 1981-DEC 1989

    Set  Items  Description
    ---  -----  -----------
?s 390
    S1      14  390
```

```
?t /5/1

 1/5/1
053989
  1989 -- The Year in Review
  December 1989,
As both the year and the decade come to a close, it's time to
recap the new databases, products, services, and...
```

2B7. Highlighting Hits

When looking at answers to a search it is useful to have the query terms in the answer highlighted so you can see why the answer was retrieved. Throughout this manual, the query terms that led to a retrieval are printed in **boldface**, but DIALOG will not do this until it is requested to do so.

Highlighting of answers can be invoked in the DIALOG search software system with the command SET HILIGHT. If the command is used in this way, the terms in the output that caused the retrieval will be displayed in **boldface**:

```
?s mp=123
        S2    17722    MP=123

?t/k/1

 2/K/1
Crystals
  Melting Point: 123 - 124 C; Solvent: CH2Cl2, diethyl ether;
  (Ref. 1

?set hi on
HILIGHT set on as ' '

?t/k/1
```

  2/K/1
Crystals
  Melting Point: **123 - 124** C; Solvent: CH2Cl2, diethyl ether;
  (Ref. 1

If your terminal will not support boldfacing of characters, then the SET command should define some character, such as * or %, and the retrieval terms will be preceded and followed by that character:

```
?set hi *
HILIGHT set on as '*'

?t/k/1


 2/K/1
Crystals
  Melting Point: *123 - 124* C; Solvent: CH2Cl2, diethyl ether;
  (Ref. 1
```

2B8. Searching and Expanding Search Terms

In the DIALOG search software system you initiate a search with the command word SELECT (abbreviated S) followed by the term you wish to search. You may search for a specific item (e.g., ethyl) or, as described in more detail in section 2B10 of this chapter, you may use truncation (e.g., eth?) for a more general search. If you are not sure of what term to use or the exact spelling of the term you may wish to use the EXPAND command (abbreviated E). This command allows you to scan the terms which are in the database. You can then choose (i.e., SELECT) up to 50 of the terms you see in the expand list which meet your searching criteria:

```
?e chrysanthem

Ref    Items   Index-term
E1         1   CHRYSANTHANON-SEMICARBAZON
E2         1   CHRYSANTHANYL
E3         7  *CHRYSANTHEM
E4         2   CHRYSANTHEMAL
E5         1   CHRYSANTHEMAMID
E6         2   CHRYSANTHEMAN
E7        82   CHRYSANTHEMAT
E8         5   CHRYSANTHEMAXANTHIN
E9         1   CHRYSANTHEMEUM
E10        2   CHRYSANTHEMIN
```

```
E11       1   CHRYSANTHEMINSAEUREMETHYLESTER
E12       1   CHRYSANTHEMIUM

          Enter P or E for more

?s e7
          S3      82   "CHRYSANTHEMAT"

?d /k
          Display 3/K/1

    Synonym: 2,4-Diallyl-3,5-dimethyl-1-indanyl-chrysanthemat

                           end of display
```

2B9. Search Operators

In the DIALOG search software system there are a number of search operators. They consist of *Boolean* operators, *proximity* operators, and *numeric* operators. All three are very important in searching the Beilstein database, as you will see in the next two chapters.

2B9A. Boolean Operators

The DIALOG search software system supports the usual three Boolean operators:

AND   OR   NOT

For the general search command shown below the Boolean operators mean the following:

A AND *B*

retrieves all Beilstein records in which the search term *A* and the search term *B* are found;

A OR *B*

retrieves all Beilstein records in which the search term *A* or the search term *B* is found;

A NOT *B*

retrieves Beilstein records in which the search term *A* is found, but not the search term *B*.

2B9B. Proximity Operators

The DIALOG search software system supports a number of proximity search operators. These are similar to the Boolean AND operator except they apply in a much more narrow and more well-defined environment within a given Beilstein record. The definition of the proximity operators, in descending order of discrimination, is:

Code ·	Explanation
W(ith)	Neighboring (adjacent) terms in the indicated order
*n*W	Neighboring terms in the indicated order which are found within *n* intervening words
N(ear)	Neighboring (adjacent) terms in either order
*n*N	Neighboring terms in any order which are found within *n* intervening words
L(ink)	Terms found in the same descriptor (e.g., NMR spectral field)
S(ubfield)	Terms found in the same subfield (e.g., the solvent for a UV spectrum associated with a specific absorption, and not the absorption from another reference)
F(ield)	Terms found in the same field but not necessarily in the same subfield

A proximity search operator must be surrounded by parentheses in order for the DIALOG software to understand that it is a proximity operator and not part of a search term.

In the search below, the single retrieval has the two search words in the order stated, within two words of one another:

```
?s (imidazole(2w)methyl)/cn
         10026   IMIDAZOLE/CN  (see also imidazol)
        212877   METHYL/CN  (see also methy)
    S4       1   (IMIDAZOLE(2W)METHYL)/CN

?t/k
 4/K/1
4,4'-(1H-imidazole-4,5-diyldimethyl)-bis-phenol
```

If the search terms are reversed, however, 1434 hits are obtained:

```
?s (methyl(2w)imidazole)/cn
         212877  METHYL/CN  (see also methy)
          10026  IMIDAZOLE/CN  (see also imidazol)
    S5      1434  (METHYL(2W)IMIDAZOLE)/CN

?t/k

 5/K/1
4-nitro-2,6-bis-trifluoromethyl-benzoimidazole-1-carboxylic acid
4-nitro-phenyl ester
```

Use of the 2n operator retrieves 1434 + 1 = 1435 hits:

```
?s (methyl(2n)imidazole)/cn
         212877  METHYL/CN  (see also methy)
          10026  IMIDAZOLE/CN  (see also imidazol)
    S6      1435  (METHYL(2N)IMIDAZOLE)/CN
```

and use of the less discriminatory (S) or (F) operators leads to many more hits:

```
?s (methyl(s)imidazole)/cn
         212877  METHYL/CN  (see also methy)
          10026  IMIDAZOLE/CN  (see also imidazol)
    S7      4744  (METHYL(S)IMIDAZOLE)/CN

?s (methyl(f)imidazole)/cn
         212877  METHYL/CN  (see also methy)
          10026  IMIDAZOLE/CN  (see also imidazol)
    S8      4744  (METHYL(F)IMIDAZOLE)/CN
```

2B9C. Numeric Operators

The DIALOG search software system supports a number of numeric search operators. These operators are particularly important and useful in searching the Beilstein Database, as there are vast amounts of numeric data in the Beilstein Factual File. The definition of the numeric operators are:

Code	Explanation
=	Equal to in value
<	Less than in value
>	Greater than in value
=< or <=	Less than or equal to in value
=> or >=	Greater than or equal to in value
:	Within a range of values

2B9D. Priority of Execution of Search Operators

The Boolean, proximity, and numeric operators are executed in the following ordered list of six classes, with the numeric operators being executed first. If there are two operators in the same class, the order of precedence is left to right in the order in which the operators are encountered. The use of parentheses can change the order, as all items within parentheses are executed first before proceeding to the next item in the search expression.

Order	Operator(s)
1	All numeric operators
2	Parentheses
3	Proximity operators
4	AND
5	OR
6	NOT

2B10. Truncation of Search Terms

The DIALOG search system software allows only a single symbol for truncation, the ? mark. It can be used in several ways.

A single ? following the word stem permits open truncation. Any number of characters may follow the stem. If *n* ? signs are embedded in a word, then exactly *n* letters will be allowed. Thus C???H will allow COACH and CATCH, but not CLUTCH. A ? followed by one space and a second ? will allow only one additional character after the first ?. Thus CAT? ? retrieves CAT and CATS, while CAT? will retrieve longer words, such as CATEGORY.

2B11. Range Searching

The DIALOG search system software allows for numeric range searching. As there are many fields in the Beilstein Database with numeric data, the reader should refer to the field of interest in the next chapter of this manual to see exactly how to search for a numeric value using the range search capability as it applies to that field.

A numeric range can be established within a search statement by means of inequalities or with the colon (:). When searching for a melting point between 100 and 110, for example, the search statements s 100<=mp<=110 and s mp=100:110 are equivalent and both correct. The search statements 100<mp<110 will allow melting points between 100.01 and 109.99 only.

The use of < or > is not allowed in all circumstances, but the colon is universally permitted. Accordingly the colon is used predominantly throughout this manual.

2B12. Printing Answers

The result of any search is a list of Beilstein records. The number of records in any such list is provided to you when the list is created. In order to see the actual records and the information and data on the chemicals which are answers to your query, it is necessary to display or type or print out the record. This is done with the DISPLAY, TYPE, and PRINT commands. You can display or type the information online as soon as the search is completed, or you can wait to see all the chemical records by asking for an offline printing to be mailed to you. The following two sections describe these alternatives.

2B12A. Online Display (Print)

The result of a search is an S# with a specific number of answers. You can display online all or part of any S#. In addition you can also display the record for a given chemical if you know its Beilstein Registry Number (BN). The command for online display is:

DISPLAY S#/format/n-m

where *n* and *m* are numbers within the range of answers from a search and "format" is a list of the fields you wish displayed (see section 2B13 below). For example, if you have 35 hits from your first search, they will be in an S1 file of 35 postings. A display of some of these answers could take the form

DISPLAY//25-28   or   D//1-3   or   d//32-35

This is discussed in more detail in section 4A.

The TYPE command is very closely related to the DISPLAY command, and its command structure is identical. It differs from DISPLAY in that it does not enter page breaks into the output and it does not pause between entries as DISPLAY does. Both commands are used interchangeably in this manual.

The TYPE and DISPLAY commands are discussed in more detail in Chapter 4 of this manual, and the means of format specification is described below.

2B12B. Offline Printing

The result of a search is an S# with a specific number of answers. You can print all or part of any S# offline. In addition you can also print the record for a given chemical if you know its Beilstein Registry Number (BN). The command for initiating an offline print is:

PRINT

The system then responds by asking which S# or BN you wish to have printed. You also must specify the print format you want to use. You are also asked if the name and address on file is the one to which these prints should be mailed. For further details please refer to DIALOG's *Searching DIALOG: The Complete Guide*.

2B13. Formats and Default Formats for Display and Printing

The second argument in a display command is reserved for the display format. Thus a command like d 2/xyz/3 will cause the third record of the second set to be displayed in xyz format. Many of the field names from the database can be used in the format argument; thus d 2/mw/3 will produce a display of the molecular weight for that record. Any number of fields can be adduced in a single command, so d 2/mw mf cn ln/3 is valid.

The superfields shown in the chart in Figure 1 (section 2B15) may also be used as formats. A command such as d 2/pr/3 will lead to all the preparative data from record 3 of set 2. Combinations of single fields and superfields such as d 2/pr mw mf/3 are allowed. The more important superfields are:

19	All display fields (lengthy display!)
BI	Basic Information—compound identity plus data summary
PP	All physical properties data, plus references
CR	All reactions data
PR	All preparative data
GR	Chemical structure, no stereochemistry included
GS	Chemical structure, with stereochemistry

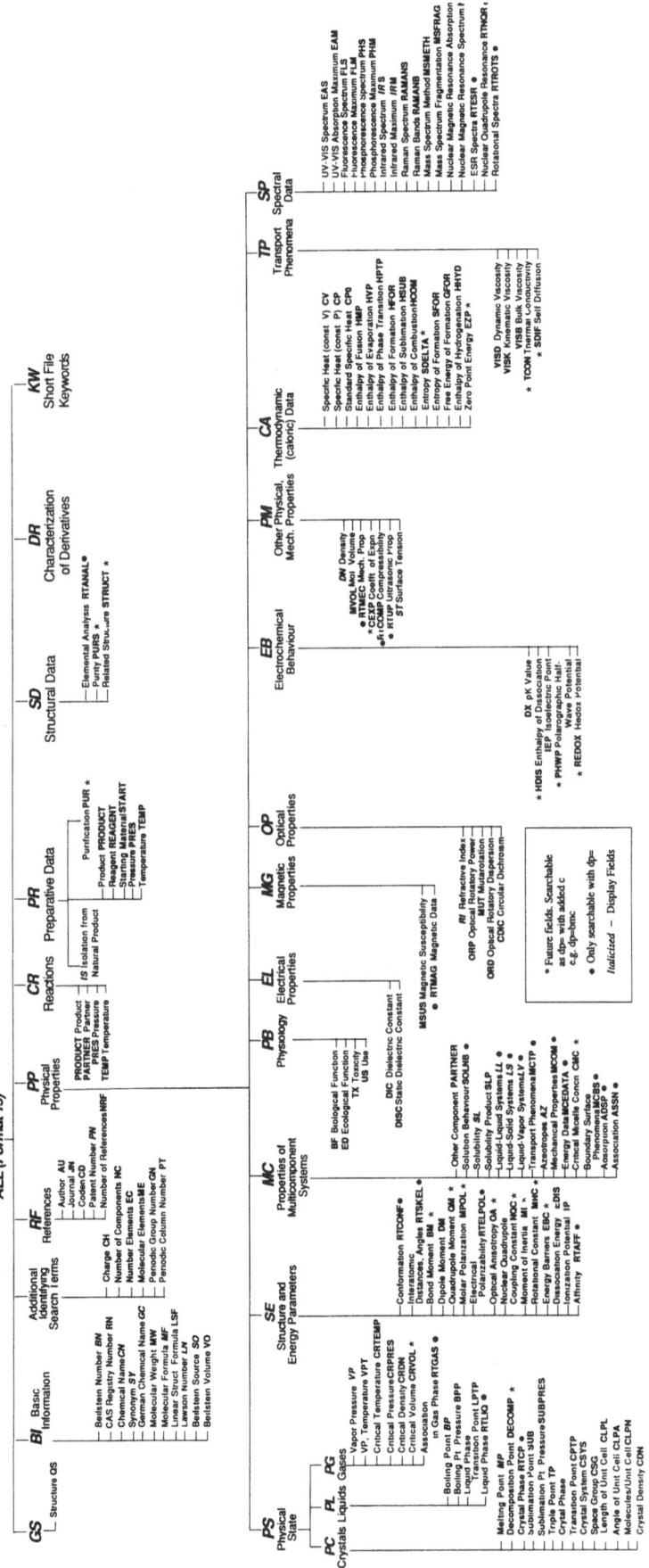

FIGURE 1.

TABLE III

FORMAT NUMBER	FIELDS SUPPLIED	CONTENT
1	BN	DIALOG Accession Number
2		Identification and Data Present
3	PR, CR	Identification, Preparations and Chemical Reactions
4	OP	Identification and Optical Properties
5	BI, SD, PR, PB, CR, PP, DR, KW	Complete Factual Record, with References at the end of the Record
6	PS	Identification, Physical State Properties
7	PP (no OP, PS) PB, DR, KW	Identification, Physical and General properties other than Physical State and Optical properties
8	SD	Identification, Structural Constituent data and Comments
9	BI, SD, PR, PB, CR, PP, DR, KW, BI, RF	Complete Factual Record with References by Section
11	BI	Basic Information and data present
12		Format 2 plus Graphic Structure (GS) plus RF
13		Format 3 plus GS, RF
14		Format 4 plus GS, RF
15		Format 5 plus GS
16		Format 6 plus GS, RF
17		Format 7 plus GS, RF
18		Format 8 plus GS, RF
19		Format 9 plus GS, RF
21		Format 11 plus GS
22		Format 2 plus GR, RF
23		Format 3 plus GR, RF
24		Format 4 plus GR, RF
25		Format 5 plus GR
26		Format 6 plus GR, RF
27		Format 7 plus GR, RF
28		Format 8 plus GR, RF
29		Format 9 plus GR, RF
31		Format 11 plus GR

The Beilstein Database has a number of default or predefined formats for display-ing or printing chemical records. Before describing these it is important to remind you that some Beilstein records are *very* long. Some are so long that the entire record will *never* be available online. This is one of the reasons why continued access to the printed Handbook is very desirable. Before you display or print any chemical record, it may be advisable to check on the chemical to see how many fields there are and how many items there are per field. This can be done with the help of format number 2. Some compounds, such as furfural or pyridine, will require hundreds of pages for the entire Beilstein record.

The predefined formats are chemically sensible collections of single fields. Thus BI (Basic Information) covers all the data relating to the identity of the compound, i.e. CN, MW, MF, LN, and so on, but not structure. The main formats are formats 1–11, which are defined in the table below. If 10 is added to a format number—i.e. format 5 becomes format 15—the new format (15) will have all the data of the old format (5), plus the chemical structure with stereochemistry and all the references at the end of the display. Addition of 20 to the format number (5 → 25) will give all the data produced by the old format (5), together with the chemical structure (no stereochemistry) and all the references at the end of the display.

The predefined formats are summarized in Table III.

In the Beilstein File the default display format is format (providing information about the identification of substance) plus those display fields in which your search terms appear. A special format (KWIC—keyword in context, abbreviated K) is available to assist in determining the reason for the retrieval of the record. This can be used alone or with other fields or fixed combinations for displaying search results.

Hit terms can be highlighted in the BI (Basic Information) display field. The KWIC format can be used in fields like PRE (Preparation) and REA (Reaction), which may contain many items. In this way, an answer set may be limited to just those subfields which contain hit terms.

2B14. Schematic Diagram of the Beilstein Database

The entire Beilstein Database is diagrammed in Figure 1.

III. Factual Data Search and Display

3A. PRELIMINARIES

The central purpose of the online Beilstein Database is to permit the locating and retrieval of selected data from this very large file. The two operations you will become familiar with if you work very long with the system are *searching* (the means of locating items), and *display* (the means of seeing the things you have retrieved).

This manual describes both these subjects, in that order and in considerable detail. The two subjects are, however, interdependent in the sense that one cannot discuss one if the other is unknown. Consequently, this introduction, which is followed by the section on searching, provides you with a thumbnail sketch of the system's search and display capabilities.

3A1. Searching of the Database

The prompt issued by the system is the question mark ?. If you wish to carry out a search, you must enter the letter s at this prompt. This letter should then be followed by the item you wish to search for, as in

?̲ ̲s̲ ̲p̲y̲r̲i̲d̲i̲n̲e̲

In general, as is explained in the next section, on searching, you would specify the field in which the word "pyridine" is to be sought. Thus the entry

?̲ ̲s̲ ̲p̲y̲r̲i̲d̲i̲n̲e̲/̲c̲n̲

will lead to a search through all the chemical names for the word "pyridine". This basic search strategy is made much more powerful by the availability of three additional techniques.

1. The system supports the usual *Boolean operators*, AND, OR, and NOT. Thus

? ̲ ̲s̲ ̲p̲y̲r̲i̲d̲i̲n̲e̲/̲c̲n̲ ̲n̲o̲t̲ ̲m̲e̲t̲h̲y̲l̲/̲c̲n̲   [or   s (pyridine not methyl)/cn]

   will retrieve all pyridines which do not also have the word "methyl" in their names. Boolean operators are explained in detail in section 2B9A.

2. *Proximity operators*, described in detail in section 2B9B, are also allowed. The command

? ̲ ̲s̲ ̲p̲y̲r̲i̲d̲i̲n̲e̲/̲c̲n̲(̲n̲)̲m̲e̲t̲h̲y̲l̲/̲c̲n̲

   will retrieve all compounds which have the words "pyridine" and "methyl" near one another in the chemical name, e.g.,

trimethyl-pyridine-3,5-dicarboxylic acid-dianilide

Replacement of the W proximity operator with the Boolean operator AND or the S operator (words in same sentence) leads to retrieval of compounds which have "pyridine" and "methyl" somewhere in the same name, for example,

1-⟨4-(4-**methyl**-pyrimidin-2-ylsulfamoyl)-phenyl⟩-**pyridine**

3. The Beilstein Database has a large number of numeric properties, such as melting points and boiling points. It is therefore very useful to be able to search with so-called *meta-operators*, which allow for use of ranges of numbers in search statements. The expression

? s 140<=mp<=150   [or   s mp=140:150]

will find any compound whose melting point has been reported to lie between 140 and 150 °C (exclusive). The statement

? s 140<mp<150

will handle the range inclusively and will therefore give more retrievals.

These three search techniques are used throughout the section on searching and are not explained further until you reach the appropriate section later in this manual.

3A2. Display of Results

Any search in the online Beilstein will lead to a temporary file which contains the database entries which your search hit upon. This file is identified for you with an S-number, as in S1, S2, and so on. You may create up to 200 results files in a single session, but all these files will disappear when you end your session. If you want to examine the contents of a results file at your terminal, you must use the display command—usually abbreviated to D or d (the software does not distinguish between upper- and lowercase in your input). A full description of the display command will be found in section 4A, but for the time being, here are the major rules concerning the use of display:

1. The required format for the display command is:

d sn/[a b c ...]/x[-y]

where sn represents a set number, *a*, *b*, and *c* are fields or formats (predefined collections of fields), and *x* is the entry within the set that is to be displayed; *x-y* represents a range of entries. An actual display command using this format would be

d s32/cn mf mw/7-10

This command is asking to see entries 7 through 10 of results file S32 and specifying the Chemical Name (cn), Molecular Formula (mf), and Molecular Weight (mw) fields as the ones that are wanted.

2. Any of the terms in this command (except the d) can be omitted. If you do not specify the S-number, then the system will work with the most recently created results file. If you omit the format details (cn, mf, mw above) you will get the system default, which is what the system thinks is a useful format. This is basically the data identifying the compound. Finally, if you cite no entries such as 7–10, you will be provided with just the first entry from whatever file is being displayed.

3. Many specialized formats exist and are described later. For now, we mention just two of them. If you use the word KWIC (keyword in context, abbreviated as K) as a format definition, as in d 12/kwic/1–5 or d /k, for each entry that is displayed, the output will be restricted to the field which was used in the retrieval process. So if S12 resulted from a Chemical Name search, then the KWIC display will provide just the chemical name of each of the retrieved compounds. A second special format is BI. This includes all the identifying chemical information on the compound—its name, molecular formula, molecular weight, and source references.

4. A shorthand exists to define sets of fields. If, for example, the format PR is cited, this will produce all the PReparative data for each compound displayed. The "superfield" PR is not itself a field; rather it contains several actual fields.

5. The search prefix DP=X will retrieve all records which contain data in the field X. Thus DP=BP will produce all compounds for which a boiling point is recorded.

This is all you need to know in order to begin with the next section, Identity of Chemical Compounds. We are going to assume that you can remember at least the first three of these five rules, and we are going to invoke them without further comment. By the time you have worked through all the examples given, further explanation of the display command should be unnecessary, but in spite of that, further explanation there is. A long and detailed description of data output is given in section 4A of this manual.

3B. IDENTITY OF CHEMICAL COMPOUNDS

3B1. Title Compounds

The Beilstein Database is compound oriented. Each record in the database deals with a specific compound, which has been assigned a unique *Beilstein Number* (BN). When discussing a record, this compound is referred to as the *title compound*.

The title compound in a record is identified in a variety of ways. Its *structure* is in the database; it can be displayed and searched for, as described in chapters 4 and 5, respectively. The *chemical name* of the title compound is also available for search or

TABLE IV

PROPERTY	SEARCH FIELDS		DISPLAY FIELDS	
Chemical structure	QS =	19	GS	
Chemical name	/CN CN =	19	BI	CN
Synonym	/SY SY =	19	BI	SY
German chemical name	/GC /DE	19	BI	GC
Partial chemical name	/CN CN =	—	—	—
Beilstein registry number	BN =	19	BI	BN
Molecular formula	MF =	19	BI	MF
Elements	EC =	19	BI	MF
Molecular elements	ME =	19	BI	MF
Periodic table row	PG =	19	BI	MF
Periodic table column	GN =	19	BI	MF
Molecular weight	MW =	19	BI	MW
Charge	CH =	19	CH	
Linearized structural formula	LF =	19	BI	LSF
Number of components	NC =	19	NC	
Starting material	/START	19	PR	START
Reagent	/REAGENT	19	PR	REAGENT
Product	/PRODUCT	19	PR	PRODUCT
Pressure	PRES =	19	PR	
Temperature	TEMP =	19	PR	
Reaction partner	PARTNER =	19	CR	PARTNER
Isolated from natural product	/IS	IS =	19	PR
Purification	DP = PURC	19	PR	
Beilstein source	SO =	19	BI	SO
Lawson Number	LN =	19	BI	LN
Elemental analysis	DP = RTANAL	19	SD	RTANAL
Purity	DP = PURSC	19	SD	
Related structure	DP = STRUCTC	19	SD	
Chemical derivative	DR =	19	DR	

display, as are the *molecular weight*, *molecular formula*, and Beilstein Registry Number. All of these items can be used in whole or in part for searching. Thus a full chemical name can be used, or alternatively, a part of a chemical name can be searched for. The fields which contain information pertaining to the title compound are listed in Table IV.

3B2. Chemical Names in the Preparative Data

When information is available concerning the preparation of a title compound, it is added to the record of the title compound under the subheading of *preparative data*.

TABLE V

PROPERTY	SEARCH FIELDS	DISPLAY FIELDS		
Starting material	/START	19	PR	START
Reagent	/REAGENT	19	PR	REAGENT
Product	/PRODUCT	19	PR	PRODUCT
Pressure	PRES =	19	PR	
Temperature	TEMP =	19	PR	
Reaction partner	PARTNER =	19	CR	PARTNER
Isolated from natural product	/IS IS =	19	PR	
Purification	DP = PURC	19		

The title compound may have been isolated from natural materials, and if so, this fact is carried in the IS field, which is described in section 3B5H. More commonly, however, the title compound will have been reported to have been prepared from other chemicals, and in such cases the other chemicals are identified in the preparative data, which is also searchable. The fields in the preparative data include the *starting material* (any chemical which is a precursor to the title compound), the *reagent* which is used to convert the educt to the title compound, the means of *purification* of the title compound, and the *products* (and by-products) that were identified from the reaction. The search capabilities are somewhat limited in comparison with searches for title compounds. In the preparative data fields, starting materials and reagents currently can be found by name only, although some common reagents (e.g. KOH) are identified by formula. In addition, the authors of cited papers can be searched for (see section 3B7B). The fields in the preparative and reactions data are listed in Table V.

Searching in the preparations data is described in detail in Section 3B6A.

3B3. Chemical Names in the Reactions Data

The reactions undergone by the title compound are carried in the *reactions data*. Both reaction *partners* and *reagents* are carried (the distinction between these is discussed in section 3B1A) and can be searched for by name, as can the reaction *products*. The subject, or type of reaction (e.g. diazotization, ozonolysis) is also searchable.

Searching in the reactions data is described in detail in section 3B6E.

3B4. Searching for Title Compounds

Every record in the Beilstein Database is concerned with a single specific compound, which is called the *title compound*. All the attributes in the record, such as Beilstein Number, molecular weight, and so on, apply specifically to the title compound. The complete record for a compound may contain information on other compounds, such as chemicals from which the title compound can be made, and it is important to be able to conduct searches for just the title compound. Such searching is the subject of this section. Searching for compounds other than title compounds is described in section 3B6.

3B4A. Searching for Title Compound Chemical Names

Chemical names appear in a variety of places in the Beilstein Database and so are stored in different fields. This must be remembered when searching for compounds by name.

3B4B. The Basic Index

The *Basic Index* (BI) of the Beilstein Database is a master index which includes the fields listed in Table VI. These fields are indexed as segments, words, or phrases. A search within the Basic Index will be carried out for any search term which is provided with neither a suffix (e.g. /CN) or a prefix (e.g. BN=).

There is no punctuation in the Basic Index, and searches for multiword terms must be carried out with proximity operators.

TABLE VI

Fields in Basic Index
Azeotrope
Biological function
Chemical derivative
Chemical name
German chemical name
Synonym
Comment
Controlled terms
Ecological data
Element count
Isolation
Preparation (and subfields)
Reactions (and subfields)
Solvent
Toxicity
Use

The Basic Index is useful when searches are intended to be broad, as is some-times the case when precise queries cannot be framed. If, for example, information of any sort on triols or diols is sought, the following Basic Index searches may be helpful:

```
?s triol
      S1     1495   TRIOL

?s diol
      S2    12548   DIOL
```

Records that deal with a triol and a ketone can be retrieved using the Basic Index:

```
?s triol and ketone
            1495   TRIOL
            5678   KETONE
      S3      26   TRIOL AND KETONE
```

as can those concerning phenylhydrazone derivatives of ketones:

```
?s ketone and phenylhydrazone
            5678   KETONE
            8056   PHENYLHYDRAZONE
      S4     443   KETONE AND PHENYLHYDRAZONE
```

or diketones:

```
?s diketone and phenylhydrazone
              76   DIKETONE
            8056   PHENYLHYDRAZONE
      S5      14   DIKETONE AND PHENYLHYDRAZONE
```

3B4C. Chemical Name of the Title Compound

The record for each compound in the "full file" carries a chemical name for the compound in the Chemical Name field (CN). This name has been assigned to the compound by Beilstein editors and conforms to IUPAC nomenclature rules. A record may also contain a synonymous name, in the /SY field, and a German language chemical name, in the /GC field. Entries in the "short file" typically have only a synonymous name (i.e. a name that may or may not conform to IUPAC rules) in the /SY field. The synonym and the German name both use uncontrolled vocabulary, in contrast to the Chemical Name. In the Beilstein file of 1,745,686 compounds, names of compounds are distributed as follows:

Chemical Name	/CN	453,212
Synonym	/SY	599,209
German Chemical Name	/GC	461,024

These groups do not overlap completely. In the "full file", most compounds have a chemical name and few have synonyms, while in the "short file", most have synonyms and few have chemical names. The German chemical names are spread throughout both the full file and the short file.

It should also be noted that the total number of names, synonyms, and German names is 1,513,445, and thus some 200,000 compounds in the database have no name at all. If exhaustive searching is important, then searching with chemical names or synonyms is not recommended. Structure searching (see Chapter 5) should be used instead.

Searches may be conducted for any complete chemical name or any part of a chemical name. Legitimate partial chemical names are arrived at by breaking up the full chemical name whenever any of the following characters is found:

$$- \quad < \quad > \quad (\quad) \quad \text{space}$$

Thus the name 3-nitro-thiophene can be used in its entirety:

```
?s 3-nitro-thiophene/cn
     S6        1   3-NITRO-THIOPHENE/CN
```

and the result is the record for that compound, which can be displayed with the display command, D (or d; see section 3A2):

```
?d /bi gs
        Display 6/BIGS/1

110656
3-nitro-thiophene
  German Chem. Name: 3-Nitro-thiophen
  Lawson No: 16857
  Beilstein Cit: 4-17-00-00256; 2-17-00-00037; 5-17
  Molecular Formula: C4H3NO2S
  Molecular Weight: 129.13
  No. of Ref: 28
```

```
   Graphic Structure:
   '110656'
```

```
Data Present:
   Data  Ref
   +Ref/Only UDF    Data Type
   7/3        PR Preparative Data
   2/6        CR Chemical Reactions
   13/9       PP Physical Properties
   /4         KW Short File Keywords

                    - end of display -
```

In this display, the notation 7/3 means that there are 10 preparative methods for the compound. Seven of these are provided in detail, but for each of the remaining three, all that is given is a literature reference.

The chemical name in this display is printed in **boldface** because it was the search term used to retrieve the entry. This is done throughout this manual. In practice, search terms are not automatically highlighted.

Chemical name searches can also be carried out using the /DE (descriptor) suffix:

```
?s 3-nitro-thiophene/de
      S7        1   3-NITRO-THIOPHENE/DE
```

or with the CN= prefix:

```
?s cn=3-nitro-thiophene
      S8        1   CN=3-NITRO-THIOPHENE
```

Alternatively, any of the chemical name segments can be used in a search of the CN field:

```
?s 3/cn
      S9   237320   3/CN

?s nitro/cn
      S10   28233   NITRO/CN
```

```
?s thiophene/cn
      S11   4717   THIOPHENE/CN
```

and the results files can be combined. This will generally give many more retrievals because order and proximity among the terms are no longer required. The 310 compounds whose chemical names contain at least those three fragments will be retrieved in this way:

```
?c 9 and 10 and 11
          237320  9
           28233  10
            4717  11
    S12       310  7 AND 8 AND 9

?d /bi gs
      Display 12/BIGS

384614
2-(2-methyl-6-nitro-phenyl)-thiophene-3-carboxylic acid amide
  German Chem. Name: 2-(2-Methyl-6-nitro-phenyl)-thiophen-
  3-carbonsaeure-amid
  Lawson No: 19569
  Beilstein Cit: 4-18-00-04303
  Molecular Formula: C12H10N2O3S
  Molecular Weight: 262.28
  No. of Ref: 1
  Graphic Structure:
  '384614'
```

```
Data Present:
  Data  Ref
  +Ref/Only UDF    Data Type
  1/        PR Preparative Data
  2/        PP Physical Properties

                      - end of display -
```

Although 3-nitrothiophene itself will be included in this group, many other compounds, such as the one shown above, will also be retrieved. This compound is quite distinct from 3-nitrothiophene, but its chemical name does contain the three fragments that were sought.

Either the Boolean operator AND or the proximity operator (S) (word in same sentence) may be used here. In general, since there is only one chemical name in the /CN field, AND and (S) are equivalent, provided that /CN suffix is used. If the (W) operator is used, this will require that the terms it links be adjacent to one another: (W1) means the same thing; (W5) will require that the two terms be within 5 words of each other in the sentence, i.e. the chemical name.

If the field-identifying suffix /CN is omitted from the search statement, the search will be carried out in all the fields in the Basic Index, and, provided the search terms are somewhere in a compound's record, that record will be retrieved. Recall that the Basic Index contains all the preparative and reactions data and that consequently such a search will give rise to more retrievals and to more retrievals of totally unrelated compounds:

```
?s 3
      S13   584111   3

?s nitro
      S14    96803   NITRO

?s thiophene
      S15     7523   THIOPHENE

?c 13 and 14 and 15
           584111   13
            96803   14
             7523   15
      S16      831   13 AND 14 AND 15

?d /bi
      Display 16/BI/1

1805842
ethane-1,1,2,2-tetracarboxylic acid tetraethyl ester
   German Chem. Name: Aethan-1,1,2,2-tetracarbonsaeure-
   tetraaethylester
   Lawson No: 1732, 298
   Beilstein Cit: 2-02-00-00699; 0-02-00-00858; 1-02-00-00331;
   3-02-00-02076; 4-02-00-02415
```

```
Molecular Formula: C14H22O8
Molecular Weight: 318.32
No. of Ref: 72
Data Present:
 Data  Ref
 +Ref/Only UDF    Data Type
 59/      PR Preparative Data
 25/      CR Chemical Reactions
 19/1     PP Physical Properties

                    - end of display -
```

The compound that was retrieved in the above example does not have the words "nitro" or "thiophene" in its chemical name. This record was retrieved because the search was carried out in the Basic Index and, as it happened, all three of the search terms are in the name of the starting material—the compound used to prepare the title compound.

Search terms can be linked together with Boolean operators or with proximity operators. The Boolean operator AND ensures that the search terms it links will be in the same record. They may or may not be in the same field of the record. Thus the combination of the two searches shown here:

```
?s 3
    S17    584111  3

?s cyano
    S18     17004  CYANO

?c 17 and 18
           584111  17
            17004  18
    S19     10850  17 AND 18
```

produces the 10850 records which contain "3" and "cyano" somewhere in the record. These search statements can be combined into a single search as shown below:

```
?s 3 and cyano
           584111  3
            17004  CYANO
    S20     10850  3 AND CYANO
```

If the search terms are required to be in the chemical name (CN) field, the search will be somewhat more discriminatory:

```
?s 3/cn
     S21   237320   3/CN

?s cyano/cn
     S22     4346   CYANO/CN

?c 21 and 22
          237320   21
            4346   22
     S23     2218   21 AND 22
```

The set S23 contains 2218 compounds whose chemical name records contain the search terms "3" and "cyano", not necessarily adjacent to one another, as in the example shown here:

```
?d/bi gs/140
        Display 23/BIGS/140

1804748
2,5-dicyano-3-oxo-adipic acid diethyl ester
   German Chem. Name: 2,5-Dicyan-3-oxo-adipinsaeure-diaethylester
   Lawson No: 2629, 298
   Beilstein Cit: 4-03-00-01927
   Molecular Formula: C12H14N2O5
   Molecular Weight: 266.25
   No. of Ref: 1
   Graphic Structure:
   '1804748'
```

```
Data Present:
 Data  Ref
 +Ref/Only UDF    Data Type
 1/       PR Preparative Data
 2/       PP Physical Properties

                        - end of display -
```

This search, then, does not guarantee that the term "3-cyano" will always be present; other phrases, such as "2-cyano-3-hydroxy", will also be retrieved. In order to restrict the retrievals to those containing "3-cyano", a proximity operator must be used. The operator (W) will require that the two terms be adjacent to one another and so provide fewer retrievals than before:

```
?s 3/cn(w)cyano/cn
         237320  3/CN
           4346  CYANO/CN
    S24     453  3/CN(W)CYANO/CN

?d /cn/48-50
       Display 24/CN/48

3-cyano-propene-1,2,3-tricarboxylic acid triethyl ester

                              - end of record -
?
       Display 24/CN/49

2,3-dicyano-3,5-dimethyl-adipic acid diethyl ester

                              - end of record -
?
       Display 24/CN/50

3-cyano-prop-1-ene-1,1,3-tricarboxylic acid triethyl ester

                              - end of display -
```

3B4D. Other Name Fields

Compounds can be retrieved on the basis of synonyms (/SY) or their German language chemical names (/GC). These fields may be more fruitful than the Chemical Name field

(/CN), particularly for compounds in the short file, where /CN entries are almost nonexistent.

The search in the /CN field for strychnine, for example, returns two hits:

```
?s strychnine/cn
      S25       2   STRYCHNINE/CN
```

but the same search in the /SY field produces four compounds, none of which were retrieved by the /CN search:

```
?s strychnine/sy
      S26       4   STRYCHNINE/SY

?c 25 and 26
                2   25
                4   26
      S27       0   25 AND 26
```

A search for the German word "Strychnin" in the /GC field gives five hits, two of which had been retrieved by the /CN search:

```
?s strychnin/gc
      S28       5   STRYCHNIN/GC

?c 25 and 28
                2   25
                5   28
      S29       2   25 AND 28
```

but none of which had been found in the /SY search:

```
?c 26 and 28
                4   26
                5   28
      S30       0   26 AND 28
```

The pattern, which is found frequently in the Beilstein Database, is that the Chemical Name, /CN, is used throughout the full file, where there are few synonyms. The synonym, /SY, is used in the short file, where there are essentially no (IUPAC) chemical names. German chemical names, /GC, are found in both files, but they overlap mostly with the Chemical Names.

A search for "urea" in the /CN field leads to 4015 compounds, 3930 of which carried the name "Harnstoff" in the /GC field:

```
?s urea/cn
      S31    4015   UREA/CN

?s harnstoff/gc
      S32    3930   HARNSTOFF/GC

?c 31 and 32
             4015   31
             3930   32
      S33    3930   31 AND 32
```

Thus the number of ureas to be found in the entire file is 4015. In addition to these, some 75 additional compounds are retrieved by a search for "urea" in the /SY field:

```
?s urea/sy
      S34      75   UREA/SY

?c 31 or 34
             4015   31
               75   34
      S35    4090   31 OR 34
```

An example of synonymous /CN and /GC fields from this results file is shown below:

```
?d /kwic/559
        Display 35/KWIC/559

N-ethyl-N'-(1-butoxy-2,2,2-trichloro-ethyl)- urea
   German Chem. Name: N-Aethyl-N'-(1-butoxy-2,2,2-trichlor-aethyl)-
        harnstoff

                          - end of record -
```

3B4E. Related Structures

Some chemical cross-referencing can be carried out in the Beilstein Database by means of the *Related Structure* field (STRUCT). This field, which is searchable with the dp=structc statement, carries comments which serve as pointers to other compounds which are "related" to the title compound. The relationship is generally circumstantial rather than chemical. As an example, the record for a compound whose structure was corrected will contain, in its STRUCT field, a reference to the earlier, incorrect structure.

In the example shown below, the STRUCT field in the record for propylidenurea (BN = 1811712) offers a reference to BN = 1811775, which turns out to be the closely related propenylurea, with which 1811712 had been confused by the original workers:

```
?s dp=structc
     S36   91132  DP=STRUCTC  (Related Structure Comments/refs.)

?d 36/bi gs k rf/202
        Display 36/BIGSKRF/202

1811712
propyliden-urea
   German Chem. Name: Propyliden-harnstoff
   Lawson No: 1762, 707
   Beilstein Cit: 2-03-00-00049
   Molecular Formula: C4H8N2O
   Molecular Weight: 100.12
   No. of Ref: 0
   Graphic Structure:
   '1817112'
```

```
   Structural Data
      Related Structure
          Comment :  BN=1811775 .)...

                            - end of display -

?s bn=1811775
     S37       2  BN=1811775

?d 37/bi gs/1-2
        Display 37/BIGS/1

1811775
propenyl-urea
   German Chem. Name: Propenyl-harnstoff
   Lawson No: 2947, 1762
   Beilstein Cit: 2-03-00-00049
```

```
     Molecular Formula: C4H8N2O
     Molecular Weight: 100.12
     No. of Ref: 1
     Graphic Structure:
     '1811775'
```

```
     Data Present:
      Data  Ref
      +Ref/Only UDF    Data Type
         2/       PR Preparative Data
         2/       PP Physical Properties

                              - end of record -

?

       Display 37/BIGS/2

1811712
propyliden-urea
  German Chem. Name: Propyliden-harnstoff
  Lawson No: 1762, 707
  Beilstein Cit: 2-03-00-00049
  Molecular Formula: C4H8N2O
  Molecular Weight: 100.12
  No. of Ref: 0
  Graphic Structure:
  '1811712'
```

```
                      - end of display -
```

The back reference from 1811712 to 1811775 is accomplished through the STRUCT field; the forward reference from 1811775 to 1811712, by means of the Beilstein Number, which points to both compounds.

The STRUCT field will be enhanced in the future. For the present, it is searchable only with the DP=STRUCTC statement—note the additional C in the term.

3B4F. Beilstein Registry Number

Every title compound in the Beilstein Database is identified with a *Beilstein Registry Number*, which is a unique number of up to seven digits. This number represents the fastest and most reliable way to retrieve the record for a specific compound:

```
?s bn=26612
      S38        1   BN=26612

?d /bigs
        Display 38/BIGS/1

26612
N-methyl-N'-phenyl-N-tropane-3 endo-yl-thiourea
  German Chem. Name: N-Methyl-N'-phenyl-N-tropan-3-endo-yl-
  thioharnstoff
  Lawson No: 27366, 14131, 2817, 1765
  Beilstein Cit: 4-22-00-03801
  Molecular Formula: C16H23N3S
  Molecular Weight: 289.44
  No. of Ref: 1
  Graphic Structure:
  '26612'
```

```
  Data Present:
   Data  Ref
   +Ref/Only UDF    Data Type
   1/          PR Preparative Data
   1/          PP Physical Properties

                    - end of display -
```

3B4G. CAS Registry Number

A majority of the compounds in the Beilstein Database have Chemical Abstracts Service (CAS) Registry Numbers. These numbers, like the Beilstein Registry Number, are unique identifiers of the chemical and can be used to retrieve the compound from either the Beilstein or the CAS Database. They can also be used, via a MAP command, to find literature citations to the compound in the CAS bibliographic files on DIALOG.

A search on CAS Registry Number is straightforward:

```
?s rn=58-61-7
        S39        1   RN=58-61-7

?t /cn so mf mw rn

 39/BI/1
93029
adenosine
  Beilstein Cit: 4-26-00-03598; 5-26
  Molecular Formula: C10H13N5O4
  Molecular Weight: 267.24
  CAS Reg. No: 58-61-7
```

3B5. Other Data on Title Compounds

3B5A. Full Molecular Formula Search

Every record in the Beilstein Database is indexed with the *molecular formula* of the title compound, and this molecular formula may be used to retrieve the record of the title compound and any of its isomers that may be present. The molecular formula should be entered in the usual Hill notation (C, H, then alphabetical):

```
?s mf=c9h7no3
      S40      175  MF=C9H7NO3
```

and separate elements should not be separated:

```
?s c9 h7 n o3
      S41        0  C9 H7 N O3
```

If the MF= prefix is omitted, the search will be carried out in the general index and will not be useful:

```
?s c9h7no3
      S42        1  C9H7NO3
```

This particular search retrieved one false hit, a compound, $C_9H_6O_3$, whose record contained the formula $C_9H_7NO_3$ in the STRUCT (related structure) field:

```
?d /bi kwic
       Display 42/BIKWIC/1

129031
isochroman-1,4-dione
  German Chem. Name: Isochroman-1,4-dion
  Lawson No: 18413
  Beilstein Cit: 4-17-00-06159; 5-17
  Molecular Formula: C9H6O3
  Molecular Weight: 162.14
  No. of Ref: 13
  Data Present:
   Data  Ref
   +Ref/Only UDF    Data Type
        /7      PR Preparative Data
        /2      CR Chemical Reactions
        4/      PP Physical Properties
        2/      DR Characterization Derivative
        /3      KW Short File Keywords
Structural Data
   Related Structure

                              -more-
?
       Display 42/BIKWIC/1

       ...Comment: lid beschriebene Verbindung (Mp: 154-155grad)
          als Isochroman-1,4-dion-4-oxim ( C9H7NO3 ), die frueher
          (s.H 491) als 3-(Acetoxyimino-methyl)-phthalid
          beschriebene  Verbindung (Mp: 154-155grad...

                          - end of display -
```

Such molecular formula searches can be combined with other searches, for example, for chemical name fragments:

```
?s mf=c6h8s and ethyl/cn
              17  MF=C6H8S
           86350  ETHYL/CN
     S43       3  MF=C6H8S AND ETHYL/CN

?d /bi gs/1-3
       Display 43/BIGS/1
```

```
1699671
ethyl-but-1-en-3-yn-c-yl sulfide
  German Chem. Name: Aethyl-but-1-en-3-in-c-yl-sulfid
  Lawson No: 503, 301
  Beilstein Cit: 4-01-00-02302
  Molecular Formula: C6H8S
  Molecular Weight: 112.19
  No. of Ref: 5
  Graphic Structure:
  '1699671'
```

```
  Data Present:
   Data  Ref
   +Ref/Only UDF    Data Type
       2/       PR Preparative Data
       2/       CR Chemical Reactions
       9/       PP Physical Properties

                            - end of record -
?
      Display 43/BIGS/2


105402
2-ethyl-thiophene
  German Chem. Name: 2-Aethyl-thiophen
  Lawson No: 16875
  Beilstein Cit: 4-17-00-00285; 0-17-00-00039; 2-17-00-00041;
  5-17
  Molecular Formula: C6H8S
  Molecular Weight: 112.19
  No. of Ref: 57
  Graphic Structure:
  '105402'
```

```
  Data Present:
   Data  Ref
   +Ref/Only UDF    Data Type
       8/17    PR Preparative Data
       8/11    CR Chemical Reactions
```

```
        56/8    PP Physical Properties
         1/     DR Characterization Derivative
          /5    KW Short File Keywords


                                   - end of record -
?
        Display 43/BIGS/3

1483
3-ethyl-thiophene
  German Chem. Name: 3-Aethyl-thiophen
  Lawson No: 16875
  Beilstein Cit: 4-17-00-00286; 2-17-00-00041; 0-17-00-00040;
  5-17
  Molecular Formula: C6H8S
  Molecular Weight: 112.19
  No. of Ref: 13
  Graphic Structure:
  '1483'
```

```
  Data Present:
   Data  Ref
   +Ref/Only UDF    Data Type
       2/3    PR Preparative Data
       2/1    CR Chemical Reactions
      18/2    PP Physical Properties
       /2     KW Short File Keywords

                                   - end of display -
```

A combination search of this sort frequently eliminates the ambiguity that is inherent in both the separate searches.

3B5B. Partial Molecular Formula Search

It is frequently desirable to search for some part of the complete molecular formula. In the Beilstein Database, this is allowed upon an element by element basis. The EC field records the presence or absence of specific element counts, and the ME field contains the identity of the different elements in the molecular formula. With the EC field, it is possible, for example, to retrieve all compounds that contain 17 carbons:

```
?s c17/ec
       S44   92862   C17/EC

?d /kwic/100-101
       Display 44/KWIC/100

  Molecular Formula:  C17H2808

                              - end of record -

       Display 44/KWIC/101

  Molecular Formula:  C17H2808

                              - end of record -
```

The EC field can be a suffix, as above, or a prefix. When it is used as a prefix, the number must be expressed on a scale of 1000, with leading zeros added as necessary:

```
?s ec=c17
       S45         0   EC=C17

?s ec=c0017
       S46   92862   EC=C0017
```

Combinations of element counts can be used to carry out partial molecular formula searches:

```
?s (c17 and n1)/ec
         92862   C17/EC
        391897   N1/EC
     S47  22831   (C17 AND N1)/EC

?d /cn kwic/50
       Display 47/CNKWIC/50

2-(2-ethoxy-ethyl)-2-cyano-3,7-dimethyl-octanoic acid ethyl ester
  Molecular Formula:  C17H31NO3

                              - end of display -
```

The ME (Molecular Elements) field carries the identity of all the elements in each compound. The elements are listed in alphabetical order of the element symbols, and alphabetical order must be observed in queries. A search for me=cho retrieves 211,783 compounds that contain C, H, and O:

```
?s me=cho
     S48    211783    ME=CHO
```

A similar search for C, H, N, and O retrieves 708,101 compounds:

```
?s me=chno
     S49    708101    ME=CHNO
```

In this case, however, the order of elements is important; use of CHON will fail:

```
?s me=chon
     S50        0   ME=CHON
```

If internal truncation is used, as in CH?O, then any formula in which the wildcard has been replaced by an (alphabetically and locationally correct) symbol will be retrieved:

```
?s me=ch?o
     S51   717363   ME=CH?O
```

The set S51 contains more entries than S49 because S49 contains only CHNO compounds, while S51 will contain, for example, CHFO and CHIO compounds, in addition to CHNO compounds.

The term CH?O will not retrieve CHCIO compounds; CH??O must be used, and, in addition to CHCIO compounds, this will retrieve, for example, CHBNO or CHDFO compounds.

Compounds containing C, H, I, N, and O are retrieved by the search statement

```
?s me=chino
     S52    4918   ME=CHINO

?d /kwic
     Display 52/KWIC/1

  Molecular Formula:  C9H16I2N2O4

                   - end of display -
```

The **ME** and **EC** fields may be used together in searches, with useful results. As an example, the search

```
?s me=chino and i2/ec
            4918  ME=CHINO
            2284  I2/EC
     S53     855  ME=CHINO AND I2/EC
```

retrieves all the compounds containing C, H, N, O, and just two atoms of I:

```
?d /kwic
        Display 53/KWIC/1

  Molecular Formula:  C9H16I2N2O4

                              - end of display -
```

3B5C. Periodic Table Searches

It is possible to retrieve compounds that contain specific elements, identified in terms of their position within the periodic table of elements.

Transition metals are classified in terms of their periodic table position, and retrieval is possible on this basis.

Rows in the periodic table are in the PT= field. The three major rows are the three transition rows T1, T2, and T3. The elements in these rows are shown below:

- T1:   Sc, Ti, V, Cr, Mn, Fe, Co, Ni, Cu, Zn

- T2:   Y, Zr, Nb, Mo, Tc, Ru, Rh, Pd, Ag, Cd

- T3:   Hf, Ta, W, Re, Os, Ir, Pt, Au, Hg

All compounds containing elements from the third transition row, for example, are retrieved by the search statement

```
?s pt=t3
     S54        212  PT=T3

?d /kwic
        Display 54/KWIC/1

  Molecular Formula:  C58H49AuN4O4

                              - end of display -
```

All compounds containing Group 2 elements (Be, Mg, Ca, Sr, Ba, Ra) are retrieved with the search statement

```
?s gn=a2
        S55       254  GN=A2

?d /kwic/1-2
        Display 55/KWIC/1

   Molecular Formula:  C34H34MgN4O3

                              - end of record -
?
        Display 55/KWIC/2

   Molecular Formula:  C44H28MgN4

                              - end of record -
```

Combining searches for row and column will retrieve compounds containing elements from the designated row as well as elements from the designated column. Each retrieved compound will contain at least one element from the cited row and one or more from the cited column:

```
?s pt=t2 and gn=a7
            373  PT=T2
         354898  GN=A7
     S56     133  PT=T2 AND GN=A7

?d /kwic
        Display 56/KWIC/1

   Molecular Formula:  C42H46FN4ORh

                              - end of record -
```

3B5D. Molecular Weight

A key part of the identity record for every title compound in the Beilstein Database is its *molecular weight*. This field is identified by the code MW. Molecular weights in the

Beilstein file are computed to two decimal places using natural average atomic weights
($C = 12.01$), and search terms must reflect this if they are to retrieve the correct entries.
Expressed with this precision, search statements are straightforward:

```
?s mw=416.35
        S57      68  MW=416.35

?d /mf mw/1-2
        Display 57/MFMW/1

  Molecular Formula: CHCl9Si3
  Molecular Weight: 416.35

                                - end of record -
?
        Display 57/MFMW/2

  Molecular Formula: C13H20N8O8
  Molecular Weight: 416.35

                                - end of display -
```

A convenient way of retrieving all compounds with a nominal molecular weight is to
use the ranging capability of the MW search:

```
?s 416<mw<417
        S58    3712   416<MW<417
```

This search retrieves the 3712 compounds whose molecular weights lie between
416.01 and 416.99. This will of course include the two compounds cited above with
their formula weight of 416.35 as well as BRN 1692749 (below) with its formula weight
of 416.51:

```
?d /bi gs/150
        Display 58/BIGS/150

1692749
  Lawson No: 24052
  Beilstein Cit: 5-19
  Molecular Formula: C24H32O6
```

```
Molecular Weight: 416.51
No. of Ref: 1
Graphic Structure:
'1692749'
```

```
Data Present:
 Data  Ref
 +Ref/Only UDF    Data Type
     /1    PR Preparative Data
     1/    PP Physical Properties

                    - end of display -
```

If it is necessary to search the molecular weight range inclusively, less than or equals signs (<=) must be used in the search statement:

```
?s 416<=MW<=417
     S59    3718  416<=MW<=417
```

and in this case 6 records will be retrieved in addition to the 3712 in S58.

Use of an integer or of one decimal place will usually fail unless ranging or inequalities are used:

```
?s mw=112
     S60     0  MW=112

?s mw=112.1
     S61     2  MW=112.1
```

because the molecular weights implied here (112.00 and 112.10, respectively) are not in the database. This will almost always be the case if the molecular weight is cited to less than 2 decimal places. Molecular weights expressed to two significant figures, on the other hand, cause no problem:

```
?s mw=112.19
      S62     32  MW=112.19

?d /cn mf mw/7-8
      Display 62/CNMFMW/7

ethyl-but-1-en-3-yn-c-yl sulfide
  Molecular Formula: C6H8S
  Molecular Weight: 112.19

                                 - end of record -
?
      Display 62/CNMFMW/8

  Molecular Formula: C7H14N
  Linear search formula: (C7H14N)+
  Molecular Weight: 112.19

                                 - end of record -
```

Any range of molecular weights is allowed, and this can be used to establish the distribution through the database of different molecular weight ranges:

```
?s 0<mw<100
      S63    5551   0<MW<100

?s 99<mw<150
      S64   58408   99<MW<150
```

3B5E. Charge

For any molecule which contains charged entities, the charge is carried in the CH field of the Beilstein Database. This field can be searched in the usual way; the sign of the charge should be cited, but if it is not, a positive charge will be assumed. Note the use of quotation marks, which is advisable when a nonalphanumeric character such as + or − appears in the search string:

```
?s ch=1
      S65   13810   CH=1
```

```
?s ch="+1"
        S66   13810   CH="+1"

?s ch="-1"
        S67     967   CH="-1"

?s ch="-2"
        S68      79   CH="-2"

?s ch=2
        S69    1189   CH=2

?s ch=3
        S70      17   CH=3

?d /bigs
        Display 70/BIGS/1


1669214
   Lawson No: 24225
   Beilstein Cit: 5-20
   Molecular Formula: C15H18B3N6
   Linear search formula: (C15H18B3N6)3+
   Molecular Weight: 314.78
   No. of Charges: 3
   No. of Ref: 1
   Graphic Structure:
   '1669214'
```

```
   Data Present:
    Data  Ref
    +Ref/Only UDF    Data Type
        /1    KW Short File Keywords


                        - end of display -
```

It is not possible to search for uncharged molecules (i.e. CH=0). If it is wished to eliminate all charged species from a retrieval set, this can be done with NOT logic:

```
?s pyridin?/sy not dp=ch
        21295  PYRIDIN?/SY
        16075  DP=CH  (Number Of Charges)
   S71  19468  PYRIDIN?/SY NOT DP=CH
```

3B5F. Linearized Structural Formula

The raw molecular formula of a compound may often be rewritten, or *linearized*, to show some structural information. This linearized structural formula is in the field LSF, which can be expanded using the DP= (data present) qualifier:

```
?e dp=lsf

Ref     Items   RT  Index-term
E1      1534     5  DP=LS (Liquid/solid Systems)
E2      1534        DP=LSC (Liquid/solid Systems Comments/refs.)
E3     17418       *DP=LSF (Linear Search Formula)
E4         0        DP=LSFD (Liquid/solid Systems Further Details)
E5      1073        DP=LSPA (Liquid/solid Systems Partner)
E6         8        DP=LSPRES (Liquid/solid Systems Pressure)
E7        14        DP=LSSOLV (Liquid/solid Systems Solvent)
E8       357        DP=LSTEMP (Liquid/solid Systems Temperature)
E9       517     8  DP=LV (Liquid/vapor Systems)
E10      517        DP=LVC (Liquid/vapor Systems Comments/refs.)
E11        0        DP=LVFD (Liquid/vapor Systems Further Details)
E12      328        DP=LVPA (Liquid/vapor Systems Partner)

        Enter P or E for more
?s e3
        S72   17418  DP="LSF"
```

Thus, some 17,400 compounds have a linearized structural formula. A typical linear formula is shown below:

```
?d /bigs/456
        Display 72/BIGS/456
```

```
1682258
  Lawson No: 24246, 2817
  Beilstein Cit: 5-20
  Molecular Formula: C8H11ClN
  Linear search formula: (C8H11ClN)+
  Molecular Weight: 156.63
  Synonym: 2-Chlor-1,4,6-trimethyl-pyridinium
  No. of Charges: 1
  No. of Ref: 2
  Graphic Structure:
  '1682258'
```

```
  Data Present:
   Data  Ref
   +Ref/Only UDF    Data Type
       /1    KW Short File Keywords

                        - end of display -
```

In a search of this field, there are two complications. First, the field mnemonic used in searching is LF, rather than LSF (the term used in display commands), and second, whenever a + or a − sign appears in a search term, the entire term must be surrounded by quotation marks. A search in this field for the $C_9H_{18}N^+$ ion can be done as follows:

```
?s lf="c9h18n)+"
      S73      24  LF="C9H18N)+"

?d /bi/12
      Display 73/BI/12

1634423
  Lawson No: 24150, 2817
  Beilstein Cit: 5-20
  Molecular Formula: C9H18N
  Linear search formula: (C9H18N)+
  Molecular Weight: 140.25
  Synonym: 1,6-Dimethyl-1-aza-bicyclo(3.2.1)octan
  No. of Charges: 1
```

```
No. of Ref: 1
Data Present:
 Data  Ref
 +Ref/Only UDF    Data Type
     /1    KW Short File Keywords

                         - end of display -
```

It is possible, by means of the LSF or MF fields, to locate deuterated and tritiated compounds:

```
?s lf=c5h10d1n
      S74       1  LF=C5H10D1N

?d /bi
        Display 74/BI/1

103273
1-deuterio-piperidine
  German Chem. Name: 1-Deuterio-piperidin
  Lawson No: 24081
  Beilstein Cit: 4-20-00-00305; 5-20
  Molecular Formula: C5H10DN
  Linear search formula: C5H10D1N
  Molecular Weight: 86.15
  No. of Ref: 18
  Data Present:
   Data  Ref
   +Ref/Only UDF    Data Type
       2/5    PR Preparative Data
       3/4    CR Chemical Reactions
       8/1    PP Physical Properties
       /4     KW Short File Keywords

                         - end of display -
```

3B5G. Number of Components

The Beilstein Database contains a number of references to salts and addition compounds. The number of species involved in such molecules is denoted by NC—a numeric searchable field. In the first release of the database, the value of NC is set to 1 for all entries. Thus a search for NC = 1 retrieves the whole file, while a search for NC = 2 results in zero hits.

3B5H. Isolation From Natural Products

Natural materials such as microorganisms and plants are a rich source of organic chemicals, and many published papers describe the isolation of specific chemicals from natural products. All chemicals in the Beilstein Database that have been reported to have been isolated from natural sources are identified by means of an entry in the IS field. Thus all compounds that have been isolated from natural sources will be retrieved by the search statement DP=IS:

```
?s dp=is
        S75    19693   DP=IS
```

This statement can be combined with other search statements to provide a filter on searches. As an example, the database contains 33 compounds which are presumed to be related to octacosanoic acid because their names contain the word "octacosanoic". Only 2 of these are natural products, however, and these can be identified in a single search:

```
?s octacosanoic/cn and dp=is
            33  OCTACOSANOIC/CN
         19693  DP=IS  (Isolation From Natural Product)
     S76     2  OCTACOSANOIC/CN AND DP=IS

?d /bi gs k/1-2
        Display 76/BIGSK/1

1716028
(2 R,4 R,6 R,8 R)-2,4,6,8-tetramethyl-octacosanoic acid
  German Chem. Name: (2 R,4 R,6 R,8 R)-2,4,6,8-Tetramethyl-
  octacosansaeure
  Lawson No: 1293
  Beilstein Cit: 4-02-00-01324
  Molecular Formula: C32H6402
  Molecular Weight: 480.86
  No. of Ref: 7
  Graphic Structure:
  '1716028'
```

```
   Data Present:
    Data  Ref
    +Ref/Only UDF    Data Type
       1/      PR Preparative Data
       3/      PP Physical Properties
2 R,4 R,6 R,8 R)-2,4,6,8-tetramethyl- octacosanoic  acid
Preparative Data
    Isolation from Natural Product
       Isolierung  aus dem Wachs von Tuberkelbazillen: (Ref. 5
       handbook), (Ref. 7 handbook


                          - end of record -
?
       Display 76/BIGSK/2


1714995
(S)-26-methyl-octacosanoic acid
  German Chem. Name: (S)-26-Methyl-octacosansaeure
  Lawson No: 1295
  Beilstein Cit: 4-02-00-01320
  Molecular Formula: C29H58O2
  Molecular Weight: 438.78
  No. of Ref: 2
  Graphic Structure:
  '1714995'
```

```
   Data Present:
    Data  Ref
    +Ref/Only UDF    Data Type
       2/      PR Preparative Data
       3/      PP Physical Properties
S)-26-methyl- octacosanoic  acid
Preparative Data
    Isolation from Natural Product
       Isolierung  aus dem Hydrolysat von Wollfett: (Ref. 1
       handbook)


                          - end of display -
```

3B5l. Lawson Number

The *Lawson Number* (LN) is a chemical fragment code with values currently between 0 and 32767. Each fully defined structure in the database is characterized in a selective but nonunique manner by one or more Lawson Numbers. Generally, a structure will have between one and five Lawson Numbers; three is a typical number. For example, the structure having the Beilstein Registry Number 62337 has four Lawson Numbers, as shown below:

```
?s bn=62337
        S77        1   BN=62337

?d /cn ln
        Display 77/CNLN/1

02, 03, 04-triacetyl-01-(2,2,2-trichloro-1,1-dimethyl-ethyl)-beta
-D-gluco-pyranuronic acid methyl ester
  Lawson No: 19808, 1155, 312, 289

                            - end of display -
```

Each of these Lawson Numbers refers to a separate fragment in the structure. Any fragment will always have the same Lawson Number, no matter what structure it is in, but a single Lawson Number may refer to more than one fragment.

Compounds carrying specific Lawson Numbers can be retrieved easily:

```
?s ln=19763
        S78      248   LN=19763
```

and compounds which have been assigned specific groups of Lawson Numbers can be retrieved by means of the LNSET= field:

```
?s lnset=19763 317
        S79        2   LNSET=19763 317
```

This search retrieves compounds having only those two Lawson Numbers assigned to them. To find compounds that carry those two Lawson Numbers, with or without other Lawson Numbers, it is necessary to use OR logic:

```
?s ln=(19763 or 317)
              248   LN=19763
             3581   LN=317
        S80  3823   LN=(19763 OR 317)
```

The NLN field allows retrieval of all compounds with a given number of Lawson Numbers:

```
?s nln=1
      S81   438026   NLN=1

?s nln=2
      S82   687566   NLN=2

?s nln=3
      S83   419648   NLN=3

?s nln=4
      S84   148881   NLN=4

?s nln>=4
      S85   200446   NLN>=4
```

The members of a group of compounds that all possess a given Lawson Number tend to fall into distinct subgroups. Thus an organic molecule will have several Lawson Numbers, and each of the Lawson Numbers can indicate several subgroups, or families. The structural features that are important when the Lawson Number for the compound is computed are listed below:

1. Cyclic class, including number and type of heteroatoms.

2. Functional groups.

3. Rings and double bonds in the carbon skeleton.

4. Carbon count of the fragment skeleton.

5. Degree of branching.

6. Degree of halogen and nitro substitution.

7. Chalcogen exchange (S, Se, etc. for O).

8. Ring sizes.

For every structure, each of these eight factors is determined, and they are then coded into a single Lawson Number in a reproducible fashion.

Lawson Numbers may be used in searching, as shown below:

```
?s ln=19808
      S86   2237   LN=19808

?s ln=17564
      S87   123    LN=17564
```

It is important to recognize, however, that such searches are not substructure searches but are really similarity searches. A Lawson Number search will generally retrieve all members of the family you may be seeking as well as other families, which may or may not be of interest, but which will have some similarity to the first group. The average total retrieval for any particular Lawson Number is on the order of 0.01%—i.e. 100 hits from one million compounds. The Lawson Number search is most powerful when combined with other searches (e.g. for element type or chemical name segment), which quickly eliminates the uninteresting families.

   The Lawson Number can be used to find positional isomers—something that is not a substructure search and that cannot be done by substructure searching. A search on name segments,

```
?s (ethoxy and phenyl and indole and carboxylic)/cn
        12047   ETHOXY/CN
        90670   PHENYL/CN
         4261   INDOLE/CN
        39888   CARBOXYLIC/CN
  S88       5   (ETHOXY AND PHENYL AND INDOLE AND
                 CARBOXYLIC)/CN
```

retrieves five compounds, which are all positional isomers of the first one, BRN 331081:

```
?d /bi gs
        Display 88/BIGS/1

331081
5-ethoxy-3-phenyl-indole-2-carboxylic acid-(2-diethylamino-
ethylamide)
   German Chem. Name: 5-Aethoxy-3-phenyl-indol-2-carbonsaeure
   -(2-diaethylamino-aethylamid)
   Lawson No: 26744, 3018, 2826, 298
   Beilstein Cit: 4-22-00-02375
   Molecular Formula: C23H29N3O2
   Molecular Weight: 379.5
   No. of Ref: 1
   Graphic Structure:
   '331081'
```

```
Data Present:
 Data  Ref
 +Ref/Only UDF    Data Type
    1/        PR Preparative Data
    1/        PP Physical Properties

                        - end of display -
```

If, however, a search with Lawson Numbers is carried out, 44 retrievals result:

```
?s ln=(26744 and 298)
            102  LN=26744
         133039  LN=298
    S89      44  LN=(26744 AND 298)
```

and it can be seen that extra 39 retrievals are all part of a family and should have been retrieved by the name segment search. That they were not retrieved is a sign of the inadequacy of nomenclature searching.

```
?c 89 not 88
            44  89
             5  88
    S90      39  89 NOT 88

?d /cn gs ln/30-32
        Display 90/CNGSLN/30


5-benzoyloxy-2-methyl-1-phenyl-1H-benz(g)indole-3-carboxylic acid
ethyl ester
  Graphic Structure:
  '385520'
```

```
  Lawson No: 26744, 14131, 10581, 298

                        - end of record -
```

?
            Display 90/CNGSLN/31

5-hydroxy-2-methyl-1-phenyl-1H-benz(g)indole-3-carboxylic acid
ethyl ester
  Graphic Structure:
  '385028'

    Lawson No: **26744, 14131, 298**

                              - end of record -
?
            Display 90/CNGSLN/32

1-ethyl-5-hydroxy-2-methyl-1H-benz(g)indole-3-carboxylic acid
ethyl ester
  Graphic Structure:
  '385287'

    Lawson No: **26744, 2826, 298**

                              - end of record -

Every Lawson Number has a special relationship to a further group of seven Lawson Numbers, because the Numbers exist in so-called *8-groups*. Within an 8-group, factors 1–3 (above) are constant, but factors 5–8 are changing. When factor 4 changes, this signifies a switch to a new 8-group. A Lawson Number of 289 is the first member of the 36th 8-group, because $289/8 = 36$ with a remainder of 1. A move to the next higher 8-group is equivalent to addition of a methylene unit to the fragment, and this makes browsing possible. When dealing with relatively common fragments, a move to the next 8-group is accomplished by adding 9 to the LN except in the special case where the LN divides by 8 to give a remainder of 7. In this case, add 1 to the LN. With less common fragments, one can add a methylene (move to the next 8-group) by adding 1 to the LN, unless its remainder is 7, in which case, subtract 7.

As an example, the methylthio group—a common fragment—has an LN of 292, i.e. $(36 \times 8) + 4$. To add a methylene, add 9, to get 301, which is the LN for ethylthio. To get the LN for thiopropyl, add 9 again, to get 310. In this way, a table can be built of the various homologous thioalkyl compounds:

LN	remainder	fragment	retrievals
292		S—CH3	24343
301	5	S—CH2CH3	8656
310	6	S—CH2CH2CH3	2246
319	7	S—CH2CH2CH2CH3	2159
320	0	S—CH2CH2CH2CH2CH3	469

All halogenated derivatives of a specific compound must have Lawson Numbers within the same 8-group, because moving to another 8-group implies a change in carbon content (factor 4, above). Thus it is fairly straightforward to find all such derivatives by retrieving all the compounds that have Lawson Numbers within an 8-group and eliminating all of those that do not contain halogens. Thus the search below retrieves all halogenated butyloxy compounds from the database:

```
?s (312<=ln<=319) and gn=a7
          21432   312<=LN<=319
         354898   GN=A7
    S91    3852   (312<=LN<=319) AND GN=A7


?d /cn sy ln/100-102
        Display 91/CNSYLN/100

(4-chloro-butoxy)-succinic acid bis-(4-chloro-butyl ester)
  Lawson No: 2103, 316

                          - end of record -
?
        Display 91/CNSYLN/101

malonic acid bis-(2,2,2-trichloro-1,1-dimethyl-ethyl ester)
  Lawson No: 1525, 312

                          - end of record -
?
        Display 91/CNSYLN/102

2-chloro-3-diisobutoxyphosphoryloxy-crotonic acid ethyl ester
  Lawson No: 1919, 317, 298

                          - end of record -
```

TABLE VII

PROPERTY	SEARCH FIELDS	DISPLAY FIELDS
Starting material	START = /START	PR
Reagent	REAGENT = /REAGENT	PR
Product	PRODUCT = /PRODUCT	PR or CR
Pressure	PRES =	PR or CR
Temperature	TEMP =	PR or CR
Isolation from N. Prod.	/IS	PR IS
Elemental Analysis	DP = RTANAL	SD
Purity	DP = PURSC	SD
Purification	DP = PURC	SD
Related Structure	DP = STRUCTC	SD
Chemical Derivative	DR = /DR	DR

3B6. SEARCHING FOR COMPOUNDS OTHER THAN TITLE COMPOUNDS

3B6A. Chemical Names in the Preparative Data

The *Beilstein Handbook* describes published procedures for synthesizing organic compounds. Such procedures contain the chemical names of those compounds (starting materials) which are used to prepare the title compound of the record. Searchable fields concerning the preparation of compounds include the starting material field (START), the reagent (REAGENT), and products or by-products (PRODUCT). The fields that carry preparative and reactions data are listed in Table VII.

The chemical names of the starting materials can be searched using the START field. Thus 2-ethylthiophene, whose record can be retrieved with a chemical name search

```
?s 2-ethyl-thiophene/cn
      S92       1   2-ETHYL-THIOPHENE/CN
```

can be prepared from a variety of precursors, as outlined in the PR record for the compound:

```
?d /pr
      Display 92/PR/1
```

```
   Data Present:
   Data  Ref
   +Ref/Only UDF     Data Type
      8/17   PR Preparative Data
Preparative Data
   Preparation
      Starting Material: 2-iodo-thiophene, ethyl bromide
      Reagent: absolute diethyl ether, sodium (Ref. 1 handbook)
   Preparation
      Starting Material: 2-bromo-thiophene, ethyl bromide
      Reagent: sodium, absolute diethyl ether (Ref. 2 handbook)
   Preparation
      Starting Material: acetylene, pyrite
      Temp: 300 - 310 C (Ref. 3 handbook)
   Preparation
      Starting Material: 2-acetyl-thiophene
      Reagent: amalgamated zinc, hydrochloric acid (Ref. 4
      handbook)
   Preparation


                              -more-
?

   Display 92/PR/1


   Starting Material: thiophene, ethyl bromide
   Reagent: ammonia, sodium amide (Ref. 5 handbook)
Preparation
   Starting Material: thiophene, ethyl bromide
   Reagent: ammonia, sodium (Ref. 6 handbook)
Preparation
   Starting Material: 1-(2)thienyl-ethanone
   Reagent: hydrazine hydrate, potassium hydroxide (Ref. 7
   handbook), (Ref. 8 handbook),  (Ref. 9 handbook)
Preparation
   Starting Material: 1-(2)thienyl-ethanone
   Reagent: hydrogen, carbon monoxide, octacarbonyl dicobalt
   Temp: 180 C
   Pressure: 154460 - 180200 Torr (Ref. 10 handbook)
Preparation  (Ref. 11)
Preparation  (Ref. 12)
Preparation  (Ref. 13)
Preparation  (Ref. 14)
```

Note that the search term, 2-ethylthiophene, cannot be found in this listing. This is because it is in the Chemical Name field; what was listed here was just the preparative data for 2-ethylthiophene, which is the compound formed when 2-iodothiophene, for example, reacts with ethyl bromide.

Any one of the starting materials, 2-iodothiophene, ethyl bromide, 2-bromo-thiophene, etc., can be used to retrieve this record, amongst others:

```
?s 2(w)iodo(w)thiophene/start
         209923   2/START
           4812   IODO/START
           5742   THIOPHENE/START
   S93       43   2(W)IODO(W)THIOPHENE/START
```

Entry 36 in this file, S92, is the entry for 2-ethylthiophene:

```
?d /cn kwic/36
        Display 93/CNKWIC/36

2-ethyl-thiophene
Preparative Data
    Preparation
        Starting Material: 2-iodo-thiophene , ethyl   bromide
        Reagent: absolute diethyl ether, sodium (Ref. 1 handbook...
            Preparation
        Starting Material: 2-bromo-thiophene, ethyl   bromide
        Reagent: sodium, absolute diethyl ether (Ref. 2 handbook...
            Preparation
        Starting Material: thiophene, ethyl   bromide
        Reagent: ammonia, sodium amide (Ref. 5 handbook...
Preparation
        Starting Material: thiophene, ethyl   bromide
        Reagent: ammonia, sodium (Ref. 6 handbook

                        - end of display -
```

If both the precursor chemicals, e.g. 2-iodothiophene and ethyl bromide, are used simultaneously as search terms, far fewer retrievals result:

```
?s ((ethyl(w)bromide) and (2(w)iodo(w)thiophene))/start
          88948   ETHYL/START
          16788   BROMIDE/START
           1082   ETHYL/START(W)BROMIDE/START
         209923   2/START
           4812   IODO/START
```

```
           5742   THIOPHENE/START
             43   2/START(W)IODO/START(W)THIOPHENE/START
   S94        1   ((ETHYL(W)BROMIDE) AND
                  (2(W)IODO(W)THIOPHENE))/START
```

If a chemical that is commonly used for derivatization purposes is sought in the START field, a large number of retrievals can result. Methanesulfonyl chloride is frequently used in this way:

```
?s methanesulfonyl(s)chloride/start
             909   METHANESULFONYL/START
           54722   CHLORIDE/START
   S95       435   METHANESULFONYL(S)CHLORIDE/START
```

and each of the compounds retrieved by this search can reasonably be expected to be a methanesulfonate formed by reaction of a free alcohol.

```
?d /bigs/100
        Display 95/BIGS/100

1750181
methanesulfonic acid propyl ester
  German Chem. Name: Methansulfonsaeure-propylester
  Lawson No: 2705, 307
  Beilstein Cit: 4-04-00-00013
  Molecular Formula: C4H1003S
  Molecular Weight: 138.18
  No. of Ref: 4
  Graphic Structure:
  '1750181'
```

```
  Data Present:
   Data  Ref
   +Ref/Only UDF    Data Type
      1/       PR Preparative Data
      6/       PP Physical Properties

                        - end of display -
```

In this case *n*-propanol reacts with the other starting material, methanesulfonyl chloride, as is described in the corresponding preparative data, which is displayed if the format PR is requested:

```
?d /pr/100
      Display 95/PR/100

  Data Present:
   Data  Ref
   +Ref/Only UDF    Data Type
      1/     PR Preparative Data
Preparative Data
   Preparation
      Starting Material: methanesulfonyl chloride, propan-1-ol
      Reagent: pyridine
      Temp: 0 C (Ref. 1 handbook),  (Ref. 2 handbook)
   Refs.
      1, Ross, Davis, JCSOA9, J.Chem.Soc., 1957 2420
      2, Williams, Mosher, JACSAT, J.Amer.Chem.Soc., 76 (1954)
      2984, 2985

                        - end of display -
```

Reagents (REAGENT) are distinguished from compounds that contribute structurally to the material being prepared and which are thus designated as starting materials (START). Water, for example, is occasionally identified as a starting material:

```
?s water/start
      S96    857   WATER/START
```

but it is found much more frequently as a reagent:

```
?s water/reagent
      S97  24337  WATER/REAGENT

?d /kwic/1-2
      Display 97/KWIC/1

Preparative Data
   ...KSCN, N.N'-bis-chloromethyl-N.N'-distearoyl-methylenediamine
      Reagent: benzene,  water  (Ref. 1 handbook)

                        - end of record -
```

```
?
      Display 97/KWIC/2

Preparative Data
   Preparation
      Starting Material: barium salt of/the/ mercaptomethionic acid
      Reagent: iodine,  water
      Other Conditions: man zerlegt das Bariumsalz mit H2SO4 (Ref.
      handbook)
```

Acids are found very commonly as starting materials or as reagents:

```
?s acid/start
      S98   160263   ACID/START

?s acid/reagent
      S99    77257   ACID/REAGENT
```

and it is possible to retrieve reactions in which aqueous HCl or aqueous KOH is specifically identified as reagents. Reagents may be in the database as names (hydrochloric acid) or formulas (KOH). If in doubt, it is best to search both ways:

```
?s water/reagent and hydrochloric/reagent
        24337   WATER/REAGENT
        15923   HYDROCHLORIC/REAGENT
   S100  2489   WATER/REAGENT AND HYDROCHLORIC/REAGENT

?s water/reagent and koh/reagent
        24337   WATER/REAGENT
        11041   KOH/REAGENT
   S101  1068   WATER/REAGENT AND KOH/REAGENT

?d /kwic
      Display 101/KWIC/1

Preparative Data
   Preparation
      Starting Material: methylenediurea, formalyne
      Reagent:  water , KOH (Ref. 2 handbook)

                              - end of display -
```

The combination of acetic anhydride and pyridine as reagents is common:

```
?s (acetic(w)acid(w)anhydride)/reagent and pyridine/reagent
           40176   ACETIC/REAGENT
           77257   ACID/REAGENT
            7145   ANHYDRIDE/REAGENT
            6927   ACETIC/REAGENT(W)ACID/REAGENT(W)
                   ANHYDRIDE/REAGENT
           20066   PYRIDINE/REAGENT
    S102     635   (ACETIC(W)ACID(W)ANHYDRIDE)/REAGENT AND
                   PYRIDINE/REAGENT
```

but acetic anhydride is much more frequently used without pyridine:

```
?s (acetic(w)acid(w)anhydride)/reagent not pyridine/reagent
           40176   ACETIC/REAGENT
           77257   ACID/REAGENT
            7145   ANHYDRIDE/REAGENT
            6927   ACETIC/REAGENT(W)ACID/REAGENT(W)
                   ANHYDRIDE/REAGENT
           20066   PYRIDINE/REAGENT
    S103    6292   (ACETIC(W)ACID(W)ANHYDRIDE)/REAGENT NOT
                   PYRIDINE/REAGENT

?d /kwic
        Display 103/KWIC/1

Preparative Data
    Preparation
        Starting Material: hexamethylenetetramine, trifluoroacetic
        acid
        Reagent: nitric acid, trifluoroacetic  acid - anhydride
        (Ref. 1 handbook...

                            - end of record -
```

Products, and by-products when reported, are captured in the PRODUCT field. From the example shown below, it can be seen that acetylation with acetic anhydride of the cupric nitrate complex of 3-phenyl thiophene yields, as a by-product, 2-nitro-3-phenyl-thiophene:

```
?s (acetic(w)acid(w)anhydride)/reagent and (2(w)nitro(w)3(w)
    phenyl(w)thiophene)/product
           40176   ACETIC/REAGENT
           77257   ACID/REAGENT
```

```
        7145  ANHYDRIDE/REAGENT
        6927  ACETIC/REAGENT(W)ACID/REAGENT(W)
              ANHYDRIDE/REAGENT
       46126  2/PRODUCT
       11311  NITRO/PRODUCT
       35997  3/PRODUCT
       19563  PHENYL/PRODUCT
         928  THIOPHENE/PRODUCT
           2  2/PRODUCT(W)NITRO/PRODUCT(W)3/PRODUCT(W)
              PHENYL/PRODUCT(W)THIOPHENE/PRODUCT
  S104       1  (ACETIC(W)ACID(W)ANHYDRIDE)/REAGENT AND
              (2(W)NITRO(W)3(W)PHENYL(W)THIOPHENE)/PRODUCT


?d /bi kwic
      Display 104/BIKWIC/1


383885
2-nitro-4-phenyl-thiophene
  German Chem. Name: 2-Nitro-4-phenyl-thiophen
  Lawson No: 16988
  Beilstein Cit: 4-17-00-00546; 5-17
  Molecular Formula: C10H7NO2S
  Molecular Weight: 205.23
  No. of Ref: 5
  Data Present:
   Data  Ref
   +Ref/Only UDF    Data Type
      1/      PR Preparative Data
      2/      PP Physical Properties
Preparative Data
    Preparation
        Starting Material: 3-phenyl-thiophene, copper (II)-nitrate
        Reagent: acetic   acid   anhydride
        By-product: 2-nitro-3-phenyl-thiophene  (Ref. 2 handbook),
        (Ref. 3 handbook)

                            - end of display -
```

3B6B. Purification Methods

The means by which compounds can be purified constitute important information which, when reported, is carried in the PUR field. This field is not searchable in detail, but the presence or absence of such data can easily be determined, and when a record exists, it will contain a literature reference to the reported purification methods. The PUR field should be searched with the search statement dp=purc, a clause which can be combined with other search statements:

```
?s coumarin/cn and dp=purc
          3175   COUMARIN/CN
          7761   DP=PURC  (Purification (method) Comments/refs.)
    S105    15   COUMARIN/CN AND DP=PURC


?d /bi gs k/1-2
       Display 105/BIGSK/1


257508
5,6,7-trimethoxy-4-methyl-coumarin
  German Chem. Name: 5,6,7-Trimethoxy-4-methyl-cumarin
  Lawson No: 19209, 289
  Beilstein Cit: 4-18-00-02387; 0-18-00-00170; 5-18
  Molecular Formula: C13H14O5
  Molecular Weight: 250.25
  No. of Ref: 16
  Graphic Structure:
  '257508'
```

```
  Data Present:
   Data  Ref
   +Ref/Only UDF    Data Type
      8/6    PR Preparative Data
      11/       PP Physical Properties
5,6,7-trimethoxy-4-methyl-coumarin
Preparative Data
   Purification Method
      Ref . 10...


                              - end of record -
  ?
       Display 105/BIGSK/2


255130
5,7,8-trimethoxy-4-methyl-coumarin
```

```
German Chem. Name: 5,7,8-Trimethoxy-4-methyl-cumarin
Lawson No: 19209, 289
Beilstein Cit: 4-18-00-02389; 5-18
Molecular Formula: C13H1405
Molecular Weight: 250.25
No. of Ref: 9
Graphic Structure:
'255130'
```

```
Data Present:
 Data  Ref
 +Ref/Only UDF    Data Type
    4/10   PR Preparative Data
    7/     PP Physical Properties
5,7,8-trimethoxy-4-methyl-coumarin
Preparative Data
   Purification Method
      Ref . 5...
```

```
                        - end of record -
```

3B6C. Chemical Derivatives

Chemical compounds are frequently identified by conversion to crystalline derivatives which have recorded physical constants, such as melting points. The (DR) field in the Beilstein Database contains any available information on chemical derivatives of the title compound.

This field can be searched with the DP=DR search statement:

```
?s dp=dr
     S106  40290  DP=DR  (Derivative)

?d /cn k/5
      Display 106/CNK/5
```

```
3,4-diaza-hexa-2,4-diene-1,1,2,5,6,6-hexacarbonitrile
Characterization Derivatives
    as bis-tetraethylammonium compound (mp: 137-138 degree) as
    well as bis-(benzyl-trimethyl-ammonium compound...


                              - end of record -
```

and in this way, the available derivatization data for any compound can be retrieved.

Alternatively, the DR= field can actually be used to assist in identification of compounds. In the example below, a compound with melting point 68°C and giving a 2,4-dinitrophenylhydrazone melting at 115°C is identified in a single search as 2,6-dihydroxy-2,6-dimethyloctan-6-one:

```
?s mp=68 and (dinitro(w)phenylhydrazone)/dr(s)115
          5406   MP=68
          3330   DINITRO/DR
          2410   PHENYLHYDRAZONE/DR
           480   115
            29   DINITRO/DR(W)PHENYLHYDRAZONE/DR(S)115
   S107      1   MP=68 AND (DINITRO(W)PHENYLHYDRAZONE)/DR(S)115

?d /bi gs k
      Display 107/BIGSK/1


1768264
2,6-dihydroxy-2,6-dimethyl-octan-3-one
  German Chem. Name: 2,6-Dihydroxy-2,6-dimethyl-octan-3-on
  Lawson No: 1079
  Beilstein Cit: 3-01-00-03309
  Molecular Formula: C10H20O3
  Molecular Weight: 188.27
  No. of Ref: 1
  Graphic Structure:
  '1768264'
```

```
  Data Present:
   Data  Ref
   +Ref/Only UDF    Data Type
      1/      PR Preparative Data
      1/      PP Physical Properties
      2/      DR Characterization Derivative
```

```
Crystals
   Melting Point: 68 C; Solvent: petroleum ether; (Ref. 1
   handbook)
   Refs.


                                 -more-
?

      Display 107/BIMPK/1


      1, Takei, Sakato, Ono, Bl.Inst.phys.chem.Res.Abstr.Tokyo,
         10 4, Zentralblatt: 1937 I 3496
Characterization Derivatives
   ...as 2.4- dinitro - phenylhydrazone  (mp:  115-117 degree)
   (Ref. 1 handbook)

                        - end of display -
```

3B6D. Elementary Analytical Data

When compounds are reported for the first time in the literature it is customary to record an elemental analysis as part of the structure proof. Such reported analyses are identified by means of a flag in the RTANAL field of the Beilstein Database. This field can be searched with the dp=rtanal search statement, which can be used most profitably in conjunction with another search statement.

The following example shows the retrieval of 33 selenium-containing compounds for which elemental analysis has been reported:

```
?s se1/ec and dp=rtanal
          6366   SE1/EC
          8325   DP=RTANAL  (Elementary Analysis)
   S108     33   SE1/EC AND DP=RTANAL

?d /bi k rf/10
      Display 108/BIKRF/10

1309833
  Lawson No: 17927
  Beilstein Cit: 5-17
  Molecular Formula: C9H9NOSe
  Molecular Weight: 226.14
  Synonym: Selenochromanon-(4)-oxim
  No. of Ref: 1
```

```
  Data Present:
   Data   Ref
   +Ref/Only UDF     Data Type
       1/       PP Physical Properties
Refs.
   1, Renson, BSCBAG, Bull.Soc.Chim.Belg., 73(1964)483,489
  Molecular Formula:  C9H9NOSe
Structural Data
    Elementary analysis (Ref. 1)


                        - end of display -
```

3B6E. Chemical Names in the Reactions Data

For many of the compounds in the Beilstein Database, reactions they are known to undergo under various conditions are documented. For each such reaction, the basic information consists of the reaction partner (**PARTNER**), the reaction product (**PRODUCT**), the reagent(s) used (**REAGENT**), and the subject (e.g. bromination), which will be found in the /CR field. The fields in the reactions data are listed in Table VIII.

The reaction partner, product, or reagent fields can be searched with the name of a chemical to retrieve the full record for the title compound undergoing the reaction. As an example, the search term

```
?s potassium(w)permanganate/partner
         4407   POTASSIUM/PARTNER   (K)
         2092   PERMANGANATE/PARTNER
    S109 1226   POTASSIUM(W)PERMANGANATE/PARTNER
```

TABLE VIII

PROPERTY	SEARCH FIELD	DISPLAY FIELDS	
Reaction partner	/PARTNER	19	CR
Reagent	/REAGENT	19	CR
Reaction product	/PRODUCT	19	CR
Subject	/CR	19	CR
Author of cited paper	AU =	19	CR

provides 1226 records of compounds whose reaction with potassium permanganate is recorded. A similar search,

```
?s benzoic(w)acid/product
          1317   BENZOIC/PRODUCT
         30064   ACID/PRODUCT
    S110  1317   BENZOIC(W)ACID/PRODUCT
```

leads to the 1317 cases in which benzoic acid is formed as a reaction product. If the searches are combined:

```
?s potassium(w)permanganate/partner (s) benzoic(w)acid/product
          4407   POTASSIUM/PARTNER   (K)
          2092   PERMANGANATE/PARTNER
          1317   BENZOIC/PRODUCT
         30064   ACID/PRODUCT
    S111    71   POTASSIUM(W)PERMANGANATE/PARTNER (S)
                 BENZOIC(W)ACID/PRODUCT
```

one retrieves 71 compounds which, when treated with potassium permanganate, give benzoic acid:

```
?d /cn kwic/30
        Display 111/CNKWIC/30

4-methyl-2-styryl-quinoline
Chemical Reactions
   Chemical Reaction:
      Partner: potassium permanganate, sulfuric acid
      Reaction Product: benzoic acid, 4-methyl-quinoline-
      carboxylic acid-(2) (Ref. 1 handbook

                          - end of record -
```

With a combination of searches, it is possible to pose chemically significant questions. For example, the search statement

```
?s isopropylamine/cn and diazotization/cr
           28   ISOPROPYLAMINE/CN
          140   DIAZOTIZATION/CR
    S112    1   ISOPROPYLAMINE/CN AND DIAZOTIZATION/CR
```

leads to all the available information on diazotization of isopropylamines:

```
?d /cn kwic
        Display 112/CNKWIC/1

beta-chloro-isopropylamine
beta-chloro- isopropylamine
Chemical Reactions
    Chemical Reaction:
        Partner: acid
        Further Conditions:  Diazotization
        Reaction Product: 1-chloro-propanol-(2), 2-chloro-
        propanol-(1) (Ref. 3 handbook)

                            - end of display -
```

Similarly, in the example below, a published procedure for the ozonolysis of norborn-5-enes is retrieved in a single search:

```
?s norborn-5-en?/cn and ozone/partner
            12   NORBORN-5-EN?/CN
           549   OZONE/PARTNER
    S113     1   NORBORN-5-EN?/CN AND OZONE/PARTNER

?d /cn kwic
        Display 113/CNKWIC/1

norborn-5-ene-2 endo,3 endo-dicarboxylic acid-anhydride
  Chemical Reactions
    ...Chemical Reaction:
        Partner: methyl acetate,  ozone
        Temp:  -20 C
        Further Conditions: Erwaermen des Reaktionsgemisches mit
            Peroxyessigsaeure und Essigsaeure und Erhitzen...

                            - end of display -
```

3B7. Other Identifying Data

3B7A. Source References

The location in the printed *Handbook* of every record is carried in the S0 field of the record. This reference is in the form 4-17-00-05051, which points to Supplement 4, volume 17, page 5051.

Searches in the SO field can be for complete citations:

```
?s so=4-17-00-05051
       S114      5  SO=4-17-00-05051

?d /bn mf cn mw gs so ln/1-4
       Display 114/BNMFCNMWGSSOLN/1

73460
20,24-epoxy-4,4,8,14-tetramethyl-18-nor-5
      alpha-ergostan-3-one-(2,4-dinitro-phenylhydrazone)
  Lawson No: 17975, 16437
  Beilstein Cit: 4-17-00-05051
  Molecular Formula: C37H56N4O5
  Molecular Weight: 636.87
  Graphic Structure:
  '73460'
```

```
                      - end of record -


?
       Display 114/BNMFCNMWGSSOLN/2

63260
20,24-epoxy-4,4,8,14-tetramethyl-18-nor-5 alpha-ergostan-3-one
      semicarbazone
  Lawson No: 17975, 1762
  Beilstein Cit: 4-17-00-05051
  Molecular Formula: C32H55N3O2
  Molecular Weight: 513.81
```

Graphic Structure:
'63260'

- end of record -

?

Display 114/BNMFCNMWGSSOLN/3

52758
1 xi-(2)thienyl-pent-1-en-4-yn-3-one-(2,4-dinitro-
phenylhydrazone)
  Lawson No: 17975, 16437
  Beilstein Cit: 4-17-00-05051
  Molecular Formula: C15H10N4O4S
  Molecular Weight: 342.33
  Graphic Structure:
  '52758'

- end of record -

```
?
      Display 114/BNMFCNMWGSSOLN/4

3321
1 xi-(2)thienyl-pent-1-en-4-yn-3-one
   Lawson No: 17975
   Beilstein Cit: 4-17-00-05051
    Molecular Formula: C9H6OS
    Molecular Weight: 162.21
    Graphic Structure:
    '3321'
```

```
                         - end of display -
```

Here, four of the five entries from page 5051 of volume 17 from Supplement 4 are displayed. Note that each has the same SO number, and also, interestingly, all have the same Lawson Number (17975) (see section 3B5l). This is a consequence of the organization of the *Handbook*, which ensures that similar structures, which will have similar Lawson Numbers, will be located close to each other.

   Search statements may also contain truncated SO numbers, and thus it is possible to retrieve 362,353 of the entries from Supplement 4:

```
?s so=4?
>>>File 390 processing for SO = 4? stopped at SO=4-01-00-0260
     S115   362353   SO=4?
```

or the 29,837 of the records from volume 17 of Supplement 4:

```
?s so=4-17?
>>>File 390 processing for SO = 4-17? stopped at SO=4-17-00-0240
     S116   29837   SO=4-17?
```

or the 4744 records which lie between pages 5000 and 5999 of volume 17 of Supplement 4:

```
?s so=4-17-00-05?
       S117    4744    SO=4-17-00-05?
```

The entire contents of any volume of Beilstein can be retrieved with the use of the VO= field:

```
?s vo=17
       S118  111875   VO=17
```

3B7B. Literature Citations

All the information in the Beilstein Database is accompanied by the appropriate citation to the original literature. It is possible to search within these citations themselves, using fields which contain the name of the *author* (AU=), the *journal* (JN=), or the *journal coden* (CD=). Any patents that are cited can be retrieved by means of their *patent number* (PN=). The total *number of references* associated with a record is stored in the NRF= field, which can be used to retrieve specific records on this basis.

Searching within citations can be especially useful as a means of reducing large retrieval sets. As an example, the work published by Ramirez on tetrahydropyridine carboxylic acid derivatives is quickly retrieved by a combination of an author search and a chemical name search, both of which alone gave many retrievals:

```
?s pyridine/cn and au=ramirez
          8154   PYRIDINE/CN
           253   AU=RAMIREZ
    S119     8   PYRIDINE/CN AND AU=RAMIREZ

?d /bi gs k
       Display 119/BIGSK/1

385439
2-(1-ethoxycarbonyl-2-oxo-cyclohexylmethyl)-6-oxo-1,4,5,6-
tetrahydro-pyridine-3-carboxylic acid ethyl ester
   German Chem. Name: 2-(1-Aethoxycarbonyl-2-oxo-cyclohexylmethyl)-6-
oxo-1,4,5,6-tetrahydro-pyridin-3-carbonsaeure-aethylester
   Lawson No: 27166, 298
   Beilstein Cit: 4-22-00-03317
   Molecular Formula: C18H25NO6
   Molecular Weight: 351.4
```

```
    Graphic Structure:
    '385439'
```

```
    No. of Ref: 1
    Data Present:
     Data  Ref
     +Ref/Only UDF    Data Type
        1/      PR Preparative Data
        2/      PP Physical Properties
2-(1-ethoxycarbonyl-2-oxo-cyclohexylmethyl)-6-oxo-1,4,5,6-
tetrahydro-pyridine -3-carboxylic acid ethyl ester

Refs.
    1,  Ramirez, Paul, JOCEAH, J.Org.Chem., 19 (1954) 183, 189

                            - end of display -
```

Retrieval of this record without use of the author's name would require correct entry of the full chemical name.

The journal, entered as either its full name or the appropriate coden, can be used in much the same way:

```
?s jn=j.amer.chem.soc.
      S120 139942   JN=J.AMER.CHEM.SOC.

?s cd=jacsat
      S121 139942   CD=JACSAT
```

The coden, if it is known, is more reliable because it is a standard. The abbreviation of the journal name is consistent throughout the database, but it is unpredictable.

Codens or journal names are useful when narrowing down searches from a large set to a few specific references:

```
?s porphyrin/cn and cd=jacsat
             913   PORPHYRIN/CN
          139942   CD=JACSAT
      S122     69   PORPHYRIN/CN AND CD=JACSAT

?d /bi rf/1
        Display 122/BIRF/1

383502
3,3',3'',3'''-(3,7,13,17-tetrakis-methoxycarbonylmethyl-porphyrin
-2,8,12,18-tetrayl)-tetra-propionic acid tetramethyl ester
  German Chem. Name:3,3',3'',3'''-(3,7,13,17-Tetrakis-methoxy-
  carbonyl- methyl-porphyrin-2,8,12,18-tetrayl)-tetra-
  propionsaeure- tetramethylester
  Lawson No: 30648, 289
  Beilstein Cit: 4-26-00-03135
  Molecular Formula: C48H54N4O16
  Molecular Weight: 942.97
  No. of Ref: 3
  Data Present:
   Data  Ref
   +Ref/Only UDF    Data Type
       2/       PR Preparative Data
       1/       PP Physical Properties
Refs.
   1, MacDonald, Michl, CJCHAG, Can.J.Chem., 34(1956)1768,1777
   2, MacDonald, JACSAT, J.Amer.Chem.Soc., 79(1957)2659
   3, Arsenault, et al, JACSAT, J.Amer.Chem.Soc.,
      82(1960)4384,4388

                          - end of display -
```

Patent numbers are stored in the **PN=** field. Each patent number includes the country of origin, and this should be cited:

```
?s pn=us 20252684
     S123     1   PN=US 20252684
```

Patent numbers can be truncated, and so ranges of patents may be retrieved:

```
?s pn=us 202?
      S124    174  PN=US 202?

?s pn=us 2029?
      S125     30  PN=US 2029?

?s pn=fr?
>>>File 390 processing for PN = FR? stopped at PN=FR 157134
      S126  14980  PN=FR?
```

The *number of references* associated with a record can be used in search statements, with the NRF= statement. By far the most common number of references is one, but many records have more than one reference:

```
?s nrf=1
       S127 1291200  NRF=1
?s nrf=2
       S128  266894  NRF=2
?s nrf=3
       S129   83825  NRF=3
?s nrf=4
       S130   35823  NRF=4
?s nrf=5
       S131   18598  NRF=5
?s nrf>5
       S132   47015  NRF>5
```

3C. PHYSICAL STATE

A complete description of the physical state of a compound, or its *state of aggregation*, includes properties which describe the solid phase, the liquid phase, and the gas phase, all three of which, in principle, can exist for any compound.

Various parameters are used to describe each of these phases; these parameters are stored in different fields of the database. A list of the parameters is given in Table IX, and the remainder of this section discusses searching with these parameters.

The display fields are hierarchical. Thus the display field PS (physical state) contains the lower level fields PC, PL, and PG (properties of crystals, liquids and gases, respectively), and the field PG contains the lowest level fields BP, LPTP, and RTLIQ. Lowest level fields may not be used in format statements.

TABLE IX

PROPERTY	SEARCH FIELD	DISPLAY FIELD(S)		
Vapor pressure	VP =	PS	PG	VP
Vapor pressure, temperature	VPT =	PS	PG	
Critical density	CRDN =	PS	PG	
Critical pressure	CRPRES =	PS	PG	
Critical temperature	CRTEMP =	PS	PG	
Critical volume	CRVOL =	PS	PG	
Boiling point, pressure	BPP =	PS	PL	
Boiling point	BP =	PS	PL	BP
Melting point	MP =	PS	PC	MP
Melting point, solvent	SOLVENT =	PS	PC	
Sublimation point	SP =	PS	PC	
Sublimation point, pressure	SUBPRES =	PS	PC	
Decomposition point	DECOMPC	PS	PC	
Triple point	TP =	PS	PC	
Liquid phase transition point	LPTP =	PS	PC	
Crystal property description		PS	PC	
Crystal space group	CSG =	PS	PC	
Crystal system	CSYS =	PS	PC	
Crystal phase transition pt.	CPTP =	PS	PC	
Crystal lattice dimension	CLPL =	PS	PC	
Crystal lattice angle	CLPA =	PS	PC	
Number of molecules/unit cell	CLPN =	PS	PC	

3C1. Vapor Pressure

The liquid state of any compound coexists with vapor of the compound. The liquid and the vapor are in equilibrium with each other, and the position of the equilibrium, i.e. the proportion of material that is in the vapor phase, depends upon the compound and also upon the ambient temperature.

The vapor pressure of compounds is measured in Torr and is carried in the **VP** field of the Beilstein Database. The temperature at which the vapor pressure was measured is given in degrees Celsius in the **VPT** field. Searches should use both of these fields, linked by the (S) operator:

```
?s vp=15(s)vpt=25
            25   VP=15
           193   VPT=25
      S1     3   VP=15(S)VPT=25
?d /cn kwic/2
      Display 1/CNKWIC/2
oxalic acid
```

```
Gases
   ...Vapor Pressure: 1.6 - 108 Torr; Temp: 18.2 - 82 C; (Ref.
      21 handbook...

                              - end of display -
```

When seeking the specific reference to a particular measurement in a compound record where there are many references, a second search may be the best approach. In the example shown here, methanol, a vapor pressure of 29.6 Torr at 0°C was reported in Reference 67:

```
?s bn=1098229
       S2        1   BN=1098229

?d /vp
       Display 2/VP

Gases
   Vapor Pressure: 69.92 - 200.94 Torr; Temp: 13.9 - 34.1 C;
   (Ref. 72 handbook)
   Vapor Pressure: 87.32 - 102.74 Torr; Temp: 18 - 21 C; (Ref.
   73 handbook)
   Vapor Pressure: 6281.2 - 51736 Torr; Temp: 130 - 230 C;
   (Ref. 64 handbook)
   Vapor Pressure: 10 Torr; Temp:  -16 C; (Ref. 74 handbook)
   Vapor Pressure: 50 Torr; Temp: 8 C; (Ref. 74 handbook)
   Vapor Pressure: 96.1 Torr; Temp: 20 C; (Ref. 74 handbook)
   Vapor Pressure: 89.1 Torr; Temp: 18.5 C; (Ref. 75 handbook)
   Vapor Pressure: 96.3 Torr; Temp: 20 C; (Ref. 75 handbook)
   Vapor Pressure: 160.2 Torr; Temp: 30 C; (Ref. 75 handbook)
   Vapor Pressure: 124.4 Torr; Temp: 25 C; (Ref. 76 handbook)
   Vapor Pressure:  (Comment: diagram exists.) (Ref. 77
   handbook)
   Vapor Pressure: 15.5 Torr; Temp:  -10 C; (Comment: table
   exists.) (Ref. 67 handbook)
   Vapor Pressure: 29.6 Torr; Temp: 0 C; (Comment: table
   exists.) (Ref. 67 handbook)
```

There are 314 references in this record, and rather than list them all, #67 can be isolated and printed by itself as shown below:

```
?s bn=1098229 and rf=67
              1   BN=1098229
           1401   RF=67
       S3        1   BN=1098229 AND RF=67

?d /kwic
       Display 3/KWIC/1
```

```
  1098229
Refs.
   ... 67 , Young, SPRDAP, Sci.Proc.R.Dublin Soc., 12 (1909/10),
440
                              - end of display -
```

3C2. Critical Properties

All gases can be liquefied by the application of pressure. There is however, for every substance a *critical temperature*, which is the maximum temperature at which it can be liquefied, or the temperature above which liquid cannot exist. The pressure required to achieve liquefaction at this temperature is known as the *critical pressure*, usually cited in Torr. The *critical density* is the density of the compound at the critical temperature and pressure, and the *critical volume* is the volume of 1 mol of the compound at the critical temperature and pressure.

   The critical temperature, in degrees Celsius, is stored in the CRTEMP field, the critical pressure in the CRPRES field, the critical density, in g/ml, in the CRDN field, and the critical volume, in cm^3/mol, in the CRVOL field:

```
?s .3<crdn<.4 and dp=crpres and 240<crtemp<300
              16   .3<CRDN<.4
             154   DP=CRPRES  (Critical Pressure)
              67   240<CRTEMP<300
       S4      5   .3<CRDN<.4 AND DP=CRPRES AND 240<CRTEMP<300

?d /bi pg/1-2
        Display 4/BIPG/1

1209237
1,2-dimethoxy-ethane
  German Chem. Name: 1,2-Dimethoxy-aethan
  Lawson No: 514, 289
  Beilstein Cit: 4-01-00-02376; 3-01-00-02073; 1-01-00-00244;
  0-01-00-00467; 2-01-00-00518
  Molecular Formula: C4H10O2
  Molecular Weight: 90.12
  No. of Ref: 31
  Data Present:
   Data  Ref
   +Ref/Only UDF    Data Type
      8/        PR Preparative Data
      4/        CR Chemical Reactions
     51/2       PP Physical Properties
```

```
Gases
   Critical Temp: 263 C; (Ref. 12 handbook)
   Refs.
                              -more-
?
      Display 4/BIPG/1

      12, Kobe, et al, JCEAAX, J.Chem.Eng.Data, 1(1956)50,52
   Critical Pressure: 29053 Torr; (Ref. 12 handbook)
   Refs.
      12, Kobe, et al, JCEAAX, J.Chem.Eng.Data, 1(1956)50,52
   Critical Density: 0.333 g/cm**3; (Ref. 12 handbook)
   Refs.
      12, Kobe, et al, JCEAAX, J.Chem.Eng.Data, 1(1956)50,52
   Vapor Pressure: 2478.7 - 28339 Torr; Temp: 127 - 260 C;
(Ref. 16 handbook)
   Vapor Pressure:  (Ref. 10 handbook)
   Refs.
      10, Lippert, LACHDL, Liebigs Ann.Chem., 276 (1893),171
      16, Kobe, et al, JCEAAX, J.Chem.Eng.Data, 1(1956)50,53

                              - end of record -
?
      Display 4/BIPG/2

1098229
Methanol
  German Chem. Name: Methanol
  Lawson No: 289
  Beilstein Cit: 4-01-00-01227; 3-00-00-01147; 2-01-00-00252;
3-01-00-01147; 1-00-00-00131; 0-01-00-00273
  Molecular Formula: CH4O
  Molecular Weight: 32.04
  No. of Ref: 1512
  Data Present:
   Data  Ref
   +Ref/Only UDF    Data Type
    68/       PR Preparative Data
    452/      CR Chemical Reactions
   1051/190  PP Physical Properties
     9/       DR Characterization Derivative
Gases
   Critical Temp: 239.43 C; (Ref. 64 handbook)

                              -more-
?
      Display 4/BIPG/2

   Critical Temp: 240.6 C; (Ref. 65 handbook)
   Critical Temp: 240 C; (Ref. 66 handbook),  (Ref. 67
   handbook)
```

```
Critical Temp: 240 C; (Ref. 68 handbook)
Critical Temp: 241.9 C; (Ref. 69 handbook)
Critical Temp: 240.5 C; (Ref. 70 handbook)
Critical Temp: 240.2 C; (Ref. 71 handbook)
Refs.
    64, Kay, Donham, CESCAC, Chem.Eng.Sci., 4(1955)1,6
    65, Fischer, Reichel, MIKRAJ, Mikrochemie, 31 (1943/44)
        104, CA (1945)2675
    66, Salzwedel, Ann. Phys., (5) 5 (1930), 879
    67, Young, SPRDAP, Sci.Proc.R.Dublin Soc., 12 (1909/10),
        440
    68, Young, ZEPCAC, Z.Phys.Chem.Stoechiom. Verwandt-
        schaftsl., 70 (1910), 626
    69, Schmidt, LACHDL, Liebigs Ann.Chem., 266 (1891), 287
    70, Crismer, Zentralblatt: 1904 I, 1479
    71, Centnerszwer, ZEPCAC, Z.Phys.Chem.Stoechiom.
        Verwandtschaftsl., 49 (1904), 207


                            -more-
?

    Display 4/PG/2


Critical Pressure: 60731 Torr; (Ref. 64 handbook)
Critical Pressure: 59660 Torr; (Ref. 67 handbook)
Critical Pressure: 75240 Torr; (Ref. 66 handbook)
Refs.
    64, Kay, Donham, CESCAC, Chem.Eng.Sci., 4(1955)1,6
    66, Salzwedel, Ann. Phys., (5) 5 (1930), 879
    67, Young, SPRDAP, Sci.Proc.R.Dublin Soc., 12 (1909/10),
        440
Critical Density: 0.272 g/cm**3; (Ref. 64 handbook)
Critical Density: 0.2722 g/cm**3; (Ref. 67 handbook)
Critical Density: 0.358 g/cm**3; (Ref. 66 handbook)
Refs.
    64, Kay, Donham, CESCAC, Chem.Eng.Sci., 4(1955)1,6
    66, Salzwedel, Ann. Phys., (5) 5 (1930), 879
    67, Young, SPRDAP, Sci.Proc.R.Dublin Soc., 12 (1909/10),
        440
Critical Volume:  (Comment: 3.674 cmE3/g.) (Ref. 68
    handbook)
Refs.
    68, Young, ZEPCAC, Z.Phys.Chem.Stoechiom.Verwandt-
        schaftsl., 70 (1910), 626


                            -more-
```

3C3. Boiling Point

Boiling points reported in the literature are stored in the BP field, and the pressure at which they were measured in the BPP field. Searches for boiling point are done in the same way as for melting point:

```
?s bp=180
      S5    5888   BP=180

?d /kwic/1
       Display 5/KWIC/1

Liquids
  Boiling Point:  180 - 187  C; Pressure: 3.4 Torr; (Ref. 1
handbook)

                            - end of record -
```

In the case of boiling point, however, it is important to specify the pressure in Torr (BPP) as well as the temperature, because the boiling point depends upon the pressure.

The database value for the boiling point is often itself a range; the entry will be retrieved if the search statement is a single number which falls within the database range or if the search statement is a range which overlaps with the database range. Thus if the database boiling point is 145–150, then bp=146, 144<bp<146, or bp= 143:152 will retrieve it, but bp=144 or bp=151 will not:

```
?s bp=180(s)bpp=3.4
          5888   BP=180
            24   BPP=3.4
     S6      1   BP=180(S)BPP=3.4

?d /kwic
       Display 6/KWIC/1

Liquids
  Boiling Point:  180 - 187  C; Pressure:  3.4  Torr; (Ref. 1
handbook)

                            - end of display -
```

If a boiling point is given with no pressure specification, then all occurrences of that boiling point will be retrieved, irrespective of pressure. This will obviously lead to many false drops, as illustrated below:

```
?s bp=278(s)bpp=760
              84   BP=278
           20992   BPP=760
      S7      37   BP=278(S)BPP=760
```

This search produces 37 compounds whose boiling point at 760 Torr is 278°C. If the pressure is left unspecified, a further 47 compounds (S9, below) are retrieved because they were recorded as boiling at 278°C, but at some other pressure:

```
?s bp=278
      S8      84   BP=278

?c 8 not 7
              84   8
              37   7
      S9      47   8 NOT 7

?d /kwic/1-2
      Display 9/KWIC/1

Liquids
  Boiling Point:  165 - 278  C; Pressure: 0.04 - 10 Torr; (Ref. 2
handbook)

                               - end of record -
?
      Display 9/KWIC/2

Liquids
  Boiling Point:  278 - 281  C; Pressure: 0.5 Torr; (Ref. 1
handbook)

                               - end of display -
```

3C4. Melting Point

Two fields in the Beilstein Database concern *melting points*. The field MP contains the melting point reported in the chemical literature, and the field SOLVENT= carries the identity of the solvent from which the material whose melting point is cited was crystallized. Melting points are all in degrees Celsius and are cited to 2 decimal places.

Most melting points are reported as integers (e.g. 123°C), and such a number is stored in the database as 123.00.

Searching the **MP** field can be carried out with exact values of the property, or with ranges or inequalities:

```
?s mp=123
        S10   10192   MP=123

?d /cn kwic
        Display 10/CNKWIC/1

2,2'-(2,9-diaza-decanedioyl)-di-malonic acid tetraethyl ester
Crystals
  Melting Point:  121 - 123  C; Solvent: methanol; (Ref. 1
  handbook)

                              - end of display -
```

The search term can be expressed as a range by means of inequalities:

```
?s 181.1<mp<199.9
        S11   120136   181.1<MP<199.9
```

in which case the search will be exclusive (mp = 181.1 will not be retrieved), or with the range symbol, a colon (:):

```
? s mp=181:200
        S12   132434   MP=181:200

?d /kwic/6000
        Display 12/KWIC/6000

Crystals
  Melting Point:  183 - 184  C; Solvent: ethanol; (Ref. 1)

                              - end of display -
```

in which case the search is inclusive (mp = 181 and mp = 200 will both be retrieved).

The database value for the melting point is often itself a range; the entry will be retrieved if the search statement is a single number which falls within the database range or if the search statement is a range which overlaps with the database range. Thus if the database melting point is 145–150, then mp=146, 144<mp<146, or mp= 143:152 will retrieve it, but mp=144, or mp=151 will not.

The solvent from which the material was crystallized may be specified in the search, using the field SOLVENT=, which is distinct from the Basic Index field /SOLVENT:

```
?s mp=181:200(s)solvent=methanol
        132434   MP=181 : MP=200
         64300   SOLVENT=METHANOL
    S13   6829   MP=181:200(S)SOLVENT=METHANOL

?d 13/bi gs kwic rf/2000-2001
      Display 13/BIGSKWICRF/2000

967693
   Lawson No: 31839, 1155, 289
   Beilstein Cit: 5-27
   Molecular Formula: C25H29NO7
   Molecular Weight: 455.51
   Synonym: 8-(2-Hydroxy-propyl)-13-acetoxy-2,3-methylendioxy-
   9,10-dimethoxy-dibenzo(a.g)chinolizidin
   No. of Ref: 2
   Graphic Structure:
   '967693'
```

```
   Data Present:
    Data  Ref
    +Ref/Only UDF    Data Type
       /1     PR Preparative Data
      1/2     PP Physical Properties
       /1     KW Short File Keywords
Refs.
  1, Kondo, YKKZAJ, Yakugaku Zasshi, 84(1964)146,150, CA
61(1964)3155
  2, Kondo, YKKZAJ, Yakugaku Zasshi, 84(1964)146,147, CA
61(1964)3155
      Crystals
  Melting Point:  184  C; Solvent: CHCl3,  methanol ; (Ref. 1)

                    - end of record -
```

```
?
        Display 13/BIGSKWICRF/2001

967601
   Lawson No: 16437, 7221
   Beilstein Cit: 5-15
   Molecular Formula: C25H34N4O4
   Molecular Weight: 454.57
   Synonym: 17-Norkauran-16-on-(2,4-dinitro-phenylhydrazon)
   No. of Ref: 2
   Graphic Structure:
   '967601'
```

```
   Data Present:
    Data  Ref
    +Ref/Only UDF    Data Type
       2/      PP Physical Properties
Refs.
   1, Briggs, et al, JCSOA9, J.Chem.Soc., (1963)1345,1352
   2, Briggs, et al, JCSOA9, J.Chem.Soc., (1962)1840,1845
Crystals
   ...Melting Point:  193 - 195  C; Solvent: CHCl3,  methanol ;
   (Ref. 2)

                              - end of display -
```

If it is wished to list the literature references for a record, this can be done by using the RF field in the display command:

```
?s 80<mp<85
       S14   21471  80<MP<85

?d /kwic rf
       Display 14/KWICRF/1

No. of Ref: 1
Refs.
   1, Feist, CHBEAM, Chem.Ber., 45 (1912),954
Crystals
```

```
   Melting Point:  81  C; Solvent: light petroleum; (Ref. 1
handbook)


                                - end of display -
```

3C5. Sublimation Temperature

Many solids vaporize directly at specific temperatures without passing through the liquid phase. This phenomenon is known as *sublimation* and is used to purify such materials. The temperature at which sublimation occurs, in degrees Celsius, is recorded in the SUB field of the Beilstein Database, and the pressure (in Torr) at which this was measured is in the SUBPRES field:

```
?s sub=360
      S15     10  SUB=360

?d /cn kwic/1-2
        Display 15/CNKWIC/1

4,5,6,7,4',5',6',7'-octachloro-1H,1'H-(2,2')biindolylidene-
3,3'-dione
Crystals
  Sublimation:  360  C; (Ref. 1 handbook)

                                - end of record -
```

The sublimation temperature depends upon the ambient pressure (SUBPRES), and this should be included in the search term, linked to the temperature with the (S) operator. The use of the (S) operator ensures that the temperature and the pressure will be from the same experiment and not merely in the same record:

```
?s sub=100:200(s)subpres=1:200
          1801  SUB=100 : SUB=200
           715  SUBPRES=1 : SUBPRES=200
     S16   437  SUB=100:200(S)SUBPRES=1:200

?s sub=0:100(s)subpres=0:1
           794  SUB=0 : SUB=100
          1932  SUBPRES=0 : SUBPRES=1
     S17   515  SUB=0:100(S)SUBPRES=0:1

?d /bi kwic
        Display 17/BIKWIC/1
```

```
1812099
trifluoro-methanesulfonic acid amide
  German Chem. Name: Trifluor-methansulfonsaeure-amid
  Lawson No: 1766
  Beilstein Cit: 4-03-00-00035
  Molecular Formula: CH2F3NO2S
  Molecular Weight: 149.09
  No. of Ref: 5
  Data Present:
   Data  Ref
   +Ref/Only UDF    Data Type
      3/      PR Preparative Data
      4/1     PP Physical Properties
Crystals
  Sublimation:  100  C; Pressure:  0.1  Torr; (Ref. 3 handbook)

                             - end of display -
```

This compound sublimes at 100°C at a pressure of 0.1 Torr and can thus conveniently be purified by sublimation.

3C6. Decomposition Point

Many organic compounds fail to melt sharply but are subject instead to thermal decomposition at temperatures below a melting point. This *decomposition point* is recorded in the DECOMP field in degrees Celsius. This field is presently empty, but comments on compounds which have been reported to decompose before they melt may be found in the DECOMPC field, retrieved with the "data present" (DP=) qualifier:

```
?s dp=decompc
      S18   1975  DP=DECOMPC  (Decomposition Comments/refs.)

?d /kwic
      Display 18/KWIC/1

Crystals
   Decomposition :  (Ref. 1)

                             - end of display -
```

3C7. Triple Point

The *triple point* of a compound is defined as the thermodynamic state at which three phases of a substance exist in equilibrium. In the Beilstein Database, this is recorded in the TP field in degrees Celsius:

```
?s 30<tp<50
      S19      1   30<TP<50

?d /cn kwic
        Display 19/CNKWIC/1

benzo(b)thiophene
Crystals
  Triple Point:  31.35  C; (Ref. 73 handbook)

                              - end of display -

?s tp=<0
      S20      65   TP=<0

?d /cn kwic/5
        Display 20/CNKWIC/5

undec-1-ene
Crystals
  Triple Point: - 49.16  C; (Ref. 19 handbook)

                              - end of display -
```

3C8. Liquid Phase Transition Point

Some compounds are known to undergo phase transitions in the liquid phase. The temperature at which this takes place is called the *liquid phase transition point*, and is stored, in degrees Celsius, in the LPTP field:

```
?s lptp="-65.8"
      S21      1   LPTP="-65.8"
```

Note the use of quotation marks in this search statement. Single or double quotation marks are recommended whenever a search statement contains nonalphanumeric characters such as − or +:

```
?d /kwic
        Display 21/KWIC/1

Liquids
   Transition Points of Liquid Modification: − 65.8  C (Ref. 40
handbook...

                              − end of display −

?s "−66"<lptp<"−64"
        S22        1   "−66"<LPTP<"−64"
```

3C9. Crystal Properties

The Beilstein Database carries the published crystal properties of compounds in several fields. A qualitative description of crystalline materials is in the *crystal property description* field (PC). Information in this field deals with the color and shape of crystals and whether or not they contain solvent of crystallization:

```
?s gelbe
       S23    4436   GELBE

?s gelb?(w)nadel?
          34134   GELB?
          26188   NADEL?
       S24   3805   GELB?(W)NADEL?

?d /cn kwic
        Display 24/CNKWIC/1

3-(S-ethyl-isothioureido)-2-cyano-acrylic acid ethyl ester
Crystals
   Color & Other Properties:
        Gelbe Nadeln  (Ref. 1 handbook...

                              − end of display −
```

Crystals can form in any one of 230 different space groups (CSG) belonging to any of the 14 basic lattice types (CSYS). The crystalline space group (CSG) as well as the crystal system (CSYS) is carried in the file, and this permits fairly precise searches:

```
?s dp=csg
     S25      590   DP=CSG  (Space Group)

?s dp=csys
     S26     1123   DP=CSYS  (Crystal System)

?s csg=p2 1
     S27       32   CSG=P2 1

?s csys=triclinic
     S28      113   CSYS=TRICLINIC
```

Combination of these criteria with each other and with information such as color and appearance leads to relatively few retrievals:

```
?s gelb? and csys=rhomb? and csg=p2 1 2 1 2 1
          34134   GELB?
            404   CSYS=RHOMB?
             56   CSG=P2 1 2 1 2 1
     S29       2   GELB? AND CSYS=RHOMB? AND CSG=P2 1 2 1 2 1

?d /bi gs kwic/2
        Display 29/BIGSKWIC/2

2903
cyclohepta(b)furan-2-one
  German Chem. Name: Cyclohepta(b)furan-2-on
  Lawson No: 17970
  Beilstein Cit: 4-17-00-05051; 5-17
  Molecular Formula: C9H6O2
  Molecular Weight: 146.15
  No. of Ref: 12
  Data Present:
  Graphic Structure:
  '2903'
```

```
  Data  Ref
  +Ref/Only UDF    Data Type
     3/1     PR Preparative Data
    15/      PP Physical Properties
```

```
Crystals
   Color & Other Properties:
        gelb  (Ref. 4 handbook),  (Ref. 5 handbook...
   Crystal System
        rhombic  (Ref. 5 handbook),  (Ref. 8 handbook

                                      -more-
?
        Display 29/BIKWIC/2

     Space Group
        P2 1 2 1 2 1  (Ref. 5 handbook),  (Ref. 8 handbook

                              - end of display -
```

The *crystal phase transition point*, defined as the temperature at which a crystal structure reverts spontaneously to a different crystalline modification, is measured in degrees Celsius and carried in the field CPTP, which is searchable in the usual way:

```
?s cptp=150
      S30          1   CPTP=150


?s 130<cptp<160
      S31         12   130<CPTP<160


?d /bi kwic/3
        Display 31/BIKWIC/3

741883
2-amino-2-hydroxymethyl-propane-1,3-diol
  German Chem. Name: 2-Amino-2-hydroxymethyl-propan-1,3-diol
  Lawson No: 3238
  Beilstein Cit: 3-04-00-00857; 4-04-00-01903; 0-04-00-00303
  Molecular Formula: C4H11NO3
  Molecular Weight: 121.14
  No. of Ref: 29
  Data Present:
   Data  Ref
   +Ref/Only UDF     Data Type
      7/        PR Preparative Data
      11/       CR Chemical Reactions
      20/4      PP Physical Properties
Crystals
  Transition Point Crystalline Modificat:  135  C (Comment:
Umwandlung in eine nicht mehr doppelbrechende Modifikation.)
(Ref. 6 handbook)
                              - end of display -
```

At a further level of precision, *crystal lattice parameters* are carried in the Beilstein Database. The *dimensions of the unit cell*, in angstroms, are in the CLPL field, and the definitive angle(s) of the unit cell is in the CLPA field:

```
?s clpl=16:18
     S32       65  CLPL=16:18

?s clpa=102:105(s)clpl=16:18
             44  CLPA=102 : CLPA=105
             65  CLPL=16 : CLPL=18
     S33       4  CLPA=102:105(S)CLPL=16:18

?d /bi gs kwic/3
        Display 33/BIGSKWIC/3

140979
thioxanthene-9-thione
  German Chem. Name: Thioxanthen-9-thion
  Lawson No: 18036
  Beilstein Cit: 4-17-00-05309; 2-17-00-00383; 0-17-00-00359;
5-17
  Molecular Formula: C13H8S2
  Molecular Weight: 228.33
  No. of Ref: 28
  Graphic Structure:
  '140979'
```

```
  Data Present:
   Data  Ref
   +Ref/Only UDF    Data Type
      7/6     PR Preparative Data
      5/3     CR Chemical Reactions
     18/5     PP Physical Properties
      /5     KW Short File Keywords
Crystals
   Dimensions of Unit Cell
      Beta: 104.5  deg; A:  16.26  Angstrom; B: 3.93
Angstrom; C: 8.13

                                    -more-
```

The crystal lattice parameters also include the *number of molecules in the unit cell* (CLPN):

```
?s clpa=90:100(s)clpn=4
            176  CLPA=90 : CLPA=100
            364  CLPN=4
     S34      65  CLPA=90:100(S)CLPN=4

?d /bi kwic
       Display 34/BIKWIC/1

1811300
1,6-bis-chlorocarbonyl-hexa-1,5-diene-1,3,3,4,4,6-hexacarboxylic
acid hexaethyl ester
  German Chem. Name:1,6-Bis-chlorcarbonyl-hexa-1,5-dien-
1,3,3,4,4,6-hexacarbonsaeure-hexaaethylester
  Lawson No: 1756, 298
  Beilstein Cit: 3-02-00-02101
  Molecular Formula: C26H32Cl2O14
  Molecular Weight: 639.44
  No. of Ref: 2
  Data Present:
   Data  Ref
   +Ref/Only UDF    Data Type
      1/      PR Preparative Data
      6/      PP Physical Properties
Crystals
   Dimensions of Unit Cell

                              -more-
?
      Display 34/BIKWIC/1

      Beta:  98.5  deg; A: 27.86 Angstrom; B: 8.73 Angstrom;
      C: 12.96 Angstrom; N:  4 ; (Ref. 2 handbook)

                         - end of display -
```

3D. STRUCTURAL AND ENERGY PARAMETERS

Several fields in the database contain information on structural and energy parameters. These fields are listed in Table X, and we now provide a detailed description of their use in searching.

TABLE X

PROPERTY	SEARCH FIELD	DISPLAY FIELDS	
Conformation	RTCONF	SE	RTCONF
Interatomic distances and angles	RTSKEL	SE	RTSKEL
Bond moment	BM	SE	BM
Dipole moment	DM	SE	DM
Quadrupole moment	QM	SE	QM
Molecular polarization	MPOL	SE	MPOL
Electrical polarizability	RTELPOL	SE	RTELPOL
Optical anisotropy	OA	SE	OA
Nuclear quadrupole coupling const.	NQC	SE	NQC
Moment of inertia	MI	SE	MI
Molecular rotational constant	MRC	SE	MRC
Molecular Energy	RTEMOL	SE	
Energy conformational barriers	EBC	SE	EBC
Molecular Deformation	RTMDEF	SE	
Bond dissociation energy	EDIS	SE	EDIS
Ionization potential	IP	SE	IP
Affinity	RTAFF	SE	RTAFF

3D1. Conformation

Published data concerning the conformation of molecules is stored in the RTCONF field. This can only be searched for the presence or absence of data, using the DP= qualifier:

```
?s dp=rtconf
     S1     9333   DP=RTCONF  (Conformation)

?d /cn sy k/50-60
       Display 1/CNSYK/50


1,2-bis-methylsulfanyl-ethane
Conformation
    Conformation (Comment: Note of conformation:Ueber die
    Konformation in fluessigem Zustand.) (Ref. 6 handbook)

                      - end of record -
```

This field is useful in searches for information concerning the conformation of specific molecules or families of molecules:

```
?s glucose/cn and dp=rtconf
          607  GLUCOSE/CN
         9333  DP=RTCONF  (Conformation)
    S2      2  GLUCOSE/CN AND DP=RTCONF


?d /bi gs se
        Display 2/BIGSSE/1

93798
O4-alpha-D-Glucopyranosyl-D-glucose
  German Chem. Name: O4-alpha-D-Glucopyranosyl-D-glucose
  Lawson No: 17647, 1122
  Beilstein Cit: 4-17-00-03057; 5-17
  Molecular Formula: C12H22O11
  Molecular Weight: 342.3
  General Comment: and cyclic tautomeres.
  No. of Ref: 127
  'Graphic Structure:'
  93798
```

```
  Data Present:
   Data  Ref
   +Ref/Only UDF    Data Type
      3/       PR Preparative Data
     62/6      CR Chemical Reactions
     51/9      PP Physical Properties
      2/       DR Characterization Derivative
      /2       KW Short File Keywords

                              -more-
?
        Display 2/BIGSSE/1


Conformation
    Conformation (Comment: Note of conformation:Konformation
    des kristallinen Monohydrats.) (Ref. 7 handbook)
```

```
    Conformation (Comment: Note of conformation:Konformation
in Loesung.) (Ref. 8 handbook),  (Ref. 9 handbook)
   Refs.
       7, Quigley, et al, JACSAT, J.Amer.Chem.Soc., 92 (1970)
          5834
       8, Isbell, J. Res. Bur. Stand. (A), 66 (1962) 233, 236
       9, Bentley, JACSAT, J.Amer.Chem.Soc., 81 (1959) 1952

                              - end of record -
```

3D2. Interatomic Distances and Angles

Published data pertaining to the geometry of the conformation of molecules are stored in the RTSKEL field, which is one of the fields in the structural and energy parameters area.

   The RTSKEL field can only be searched for the presence or absence of data, by means of the DP= qualifier:

```
?s dp=rtskel
       S3     6651   DP=RTSKEL  (Interatomic Distances And
Angles)
?d /k/20-21
       Display 3/K/20

Skeletal Characteristics
    Electron  distribution (Ref. 1)

                              - end of record -
?
       Display 3/K/21

Skeletal Characteristics
    Electron  distribution (Ref. 1)

                              - end of display -
```

The RTSKEL field can be very useful when searching for information about the molecular geometry of a molecule, as in the example below, where data pertaining to the structure of oxyprothepin are retrieved in a single search:

```
?s sy=oxyprothepin and dp=rtskel
               2  SY=OXYPROTHEPIN
```

```
              2900  DP=RTSKEL  (Interatomic Distances And
                    Angles)
        S4       1  SY=OXYPROTHEPIN AND DP=RTSKEL
```

?d 4/bi gs se

```
        Display 4/BIGSSE/1
```

716048
  Lawson No: 28572, 20520, 292
  Beilstein Cit: 5-24
  Molecular Formula: C22H26N2OS2
  Molecular Weight: 398.58
  Synonym: Oxyprothepin
  No. of Ref: 1
  'Graphic Structure:'
  716048

  Data Present:
   Data Ref
   +Ref/Only UDF   Data Type
       /3   PP Physical Properties
Conformation
   Conformation (Ref. 1)
   Refs. 1, Koch, Evgrard, ACBCAR, Acta Crystallogr.Sect.B,
          30(1974)2925
Skeletal Characteristics
   Interatomic distances and angles (Ref. 1)

                              -more-
?
     Display 4/BISE/1

   Refs.
     1, Koch, Evgrard, ACBCAR, Acta Crystallogr.Sect.B,
        30(1974)2925

                        - end of display -

3D3. Bond Moment

Any bond joining two dissimilar elements will be polarized to some extent as a result of the different electronegativities of the atoms. As a result, the bond will resemble a small dipole or magnet and will have a bond moment which depends upon the length of the bond and the degree of polarization. Bond moments are stored in the BM field, which currently can only be searched for comments it may contain. This is done with the search statement DP=BMC:

```
?s dp=bmc
      S5        14  DP=BMC  (Bond-moment Comments/refs.)

?d /bi k/2
      Display 5/BIK/2

1718793
hydrogen cyanide
  German Chem. Name: Cyanwasserstoff
  Lawson No: 1145
  Beilstein Cit: 3-02-00-00061; 2-02-00-00037; 0-02-00-00029;
  1-02-00-00022; 4-02-00-00050
  Molecular Formula: CHN
  Molecular Weight: 27.03
  No. of Ref: 1439
  Data Present:
   Data  Ref
   +Ref/Only UDF    Data Type
    434/       PR Preparative Data
    549/       CR Chemical Reactions
    225/89     PP Physical Properties
Electrical Moment
   Bond-moment: (Comment: C-H-Bindungsmoment und
   C=-N-Bindungsmoment (aus der Intensitaet von IR-Banden...

                       - end of display -
```

3D4. Dipole Moment

The bulk electrical properties of a molecule are usually measured as their *dipole moments*, which are carried in the DM field of the Beilstein Database. The unit of dipole moments is the debye (D), and the data are expressed to two decimal places. The *temperature* at which the dipole moment was measured, in degrees Celsius, is stored in the TEMP= field. The *method of measurement* used and the *solvent* of the experiment (SOLVENT=) are also recorded.

Searches of the DM field are carried out in the usual manner; the search statement can carry a dipole moment expressed to two decimal places:

```
?s dm=4.98
        S6       2   DM=4.98

?d /cn kwic
        Display 6/CNKWIC/1

9-chloro-3-nitro-acridine
Electrical Moment
  Dipole Moment: 4.98 Debye; Method: Dielectric cnst.; Solvent:
  benzene; (Ref. 7 handbook)

                              - end of display -
```

The dipole moment may be expressed in the search statement as a range:

```
?s 4<dm<5
        S7      258   4<DM<5
```

Greater precision will be obtained if the measurement temperature is also cited in the search statement:

```
?s dm=4:5 and temp=25
            266   DM=4 : DM=5
          48524   TEMP=25
      S8    114   DM=4:5 AND TEMP=25
```

The Boolean AND operator ensures only that the TEMP= will be somewhere in the same record as the DM. To be sure that the specific dipole moment was measured at the cited temperature, the (S) proximity operator should be used:

```
?s dm=4:5(s)temp=25
            266   DM=4 : DM=5
          91909   TEMP=25
      S9     53   DM=4:5(S)TEMP=25
```

The extra 67 retrievals in S8 were false drops because they carried dipole moments of between 4 and 5 debyes, but not measured at 25 degrees:

```
?c 8 not 9
            120  8
             53  9
     S10      67  8 NOT 9

?t /k/1-2
 10/K/1

Electrical Moment
  Dipole Moment: 4.23 Debye; Method: Dielectric cnst.; (Ref. 13, hand-
book)Solvent:
  dioxane; (Ref. 8 handbook)
Other Physical & Mechanical Properties
  Density: 0.9616 g/cm**3; Ref Temp: 4 C; Measurement Temp: 25
  C; (Ref. 6 handbook)
Optical Properties
  Refractive Index: 1.4557; Wavelength: 589 nm; Temp:  25  C;
  (Ref. 6 handbook)

                              - end of record -
?
      Display 10/K/2

Electrical Moment
  Dipole Moment: 4.32 Debye; Method: Dielectric cnst.;
Solvent:
  dioxane; (Ref. 7 handbook)
Optical Properties
  ...Refractive Index: 1.4457; Wavelength: 589 nm; Temp: 25 C;
  (Ref. 4 handbook)

                              - end of display -
```

The method of measurement and the solvent used (SOLVENT=) are both search-able:

```
?s dm=4:5 and dielectric?
            266  DM=4 : DM=5
           2549  DIELECTRIC?
     S11    256  DM=4:5 AND DIELECTRIC?
```

```
?s dp=dm (s)dielectric? and solvent=benzene
            2416   DP=DM  (Dipole Moment)
            2549   DIELECTRIC?
            2199   DP=DM(S)DIELECTRIC?
           52086   SOLVENT=BENZENE
     S12    1686   DP=DM (S)DIELECTRIC? AND SOLVENT=BENZENE


?d /k
        Display 12/K/1


Electrical Moment
   Dipole  Moment: 1.36 Debye; Method:  Dielectric  cnst.;
   Solvent: benzene ; (Ref. 4 handbook)


                             - end of display -
```

It is possible to retrieve all dipole moment measurements which were not (stated to be) carried out by the dielectric constant method, although in most of these cases, the method used is simply not cited:

```
?s dp=dm not dielectric?
            2416   DP=DM  (Dipole Moment)
            2549   DIELECTRIC?
     S13     134   DP=DM NOT DIELECTRIC?

?d /k/1-2
        Display 13/K/1

Electrical Moment
   Dipole Moment: 3.6 Debye; Solvent: benzene; (Ref. 1
   handbook)


                             - end of record -
?
        Display 13/K/2

Electrical Moment
   Dipole Moment: 1.3 Debye; (Ref. 4 handbook)


                             - end of display -
```

3D5. Quadrupole Moment

If a molecule is not spherically symmetrical, it may possess a *quadrupole moment*, whose magnitude is a measure of the asymmetry. Quadrupole moments are measured in electrostatic units (esu) and cited in units of 10^{-27} esu. They are cited in the QM field, and the comments from this field can be retrieved with the command dp=qmc:

```
?s dp=qmc
      S14       16  DP=QMC  (Quadrupole Moment Comments/refs.)

?d /cn sy k /1-5
        Display 14/CNSYK/1


  Synonym: 5'-O-(beta, beta, beta-Trichlor-ethylphosphoryl)-
  thymidylyl-(3'-5')-thymidin
Electrical Moment
   Quadrupole Moment: (Ref. 1)


                          - end of record -
```

3D6. Molecular Polarizability

When a molecule is placed in an electric field, its charge distribution may be affected and there may result a dipole which can add to any permanent dipole which may already be present. If the electrical field has an intensity F and the induced dipole has a magnitude m, then the *molecular polarizability* is defined as m/F and is measured in units of cm^3/mol. These data are stored in the MPOL field of the Beilstein Database. This field can be searched with the DP=MPOLC statement:

```
?s dp=mpolc
      S15      156  DP=MPOLC

?d /cn sy k /5-7
        Display 15/CNSYK/5

hexachloro-ethane
Electrical Polarizability
   Molar Polarization: (Comment: in Hexan und Benzol.) (Ref.
   105 handbook)

                            - end of record -
```

```
?
        Display 15/CNSYK/6

acetic acid methyl ester
Electrical Polarizability
    Molar Polarization: (Comment: in Dampf bei 307.5-482.2 K.)
    (Ref. 6 handbook)

                                - end of record -

?
        Display 15/CNSYK/7

trichloro-fluoro-methane
Electrical Polarizability
    Molar Polarization: (Comment: bei 9400 MHz.) (Ref. 41
    handbook)

                                - end of record -
```

3D7. Electrical Polarizability

The *electrical polarizability* of a molecule is a measure of its tendency to form an induced dipole when it is placed in an electric field. The electrical polarizability is defined as m/F, where m is the electrical moment of the induced dipole and F is the field strength. In the Beilstein Database, the electrical polarizability is stored in the RTELPOL field, which currently can be searched using the search statement DP= RTELPOL:

```
?s dp=rtelpol
      S16       66  DP=RTELPOL  (Electrical Polarizability)

?d /cn sy k/6-7

?
        Display 16/CNSYK/6

Electrical Polarizability
    Electrical Polarization
        Electron  polarization (Ref. 4)

                                - end of record -
```

```
?
       Display 16/CNSYK/7

  Synonym: gamma-Thiobutyrolacton
Electrical Polarizability
   Electrical Polarization
       Electron  polarization (Ref. 2)

                           - end of record -
```

This field can be useful in retrieving, for example, references to measurements of the electrical polarizability of halogen-containing compounds:

```
?s gn=a7 and dp=rtelpol
       354898  GN=A7
           66  DP=RTELPOL  (Electrical Polarizability)
   S17      9  GN=A7 AND DP=RTELPOL

?d /bi k/1-3
       Display 17/BIK/1

1421700
  Lawson No: 20871
  Beilstein Cit: 5-19
  Molecular Formula: C7H13ClO2
  Molecular Weight: 164.63
  No. of Ref: 1
  Data Present:
   Data  Ref
   +Ref/Only UDF    Data Type
       /1    PP Physical Properties
  Molecular Formula:  C7H13ClO2
Electrical Polarizability
   Electrical Polarization
       Electron  polarization (Ref. 1)

                           - end of record -


?
       Display 17/BIK/2

1421699
  Lawson No: 20871
  Beilstein Cit: 5-19
```

```
Molecular Formula: C7H13ClO2
Molecular Weight: 164.63
No. of Ref: 1
Data Present:
 Data  Ref
 +Ref/Only UDF    Data Type
     /1    PP Physical Properties
Molecular Formula:  C7H13ClO2
Electrical Polarizability
   Electrical Polarization
       Electron  polarization (Ref. 1)


                        - end of record -



?

      Display 19/BIK/3

1421587
  Lawson No: 20861
  Beilstein Cit: 5-19
  Molecular Formula: C6H11ClO2
  Molecular Weight: 150.6
  Synonym: 5c-Chlor-4r,6t-dimethyl-1,3-dioxan
  No. of Ref: 1
  Data Present:
   Data  Ref
   +Ref/Only UDF    Data Type
       /2    PP Physical Properties
  Molecular Formula:  C6H11ClO2
Electrical Polarizability
   Electrical Polarization
       Electron  polarization (Ref. 1)

                        - end of display -
```

3D8. Optical Anisotropy

Molecules which possess no chiral centers and which are, in principle, optically inactive may, under the influence of UV radiation, show some optical activity. This is because absorption of energy may lead to the creation of chiral forms of the molecule in the excited state. This property, which is termed *optical anisotropy*, is a qualitative property; it has no units and is stored in the OA field, which should be searched with the DP= qualifier:

```
?s dp=oac
      S20       100  DP=OAC  (Optical Anisotropy Comments/refs.)

?d /cn sy k/50
         Display 20/CNSYK/50

3-bromo-dibenzothiophene
Electrical Polarizability
   Optical Anisotropy
      Optical  Anisotropy:  (Ref. 4)

                                   - end of record -
```

The occurrence of induced optical anisotropy in families of compounds can be examined with combination searches:

```
?s acridine/cn and dp=oac
            2242  ACRIDINE/CN
             100  DP=OAC  (Optical Anisotropy Comments/refs.)
      S21      17  ACRIDINE/CN AND DP=OAC
?d /cn gs sy k/1-4

        Display 21/CNGSSYK/1

dibenz(a,h)acridine
dibenz(a,h) acridine
   'Graphic Structure:'
   209261
```

```
Electrical Polarizability
   Optical Anisotropy
      Optical Anisotropy:  (Ref. 11)

                                   - end of record -
?
        Display 21/CNGSSYKWIC/2
```

```
dibenz(c,h)acridine
dibenz(c,h) acridine
  'Graphic Structure:'
  209259
```

```
Electrical Polarizability
   Optical Anisotropy
      Optical Anisotropy:  (Ref. 11)

                              - end of record -


?
      Display 21/CNGSSYKWIC/3

7,9-dimethyl-benz(c)acridine
7,9-dimethyl-benz(c) acridine
  'Graphic Structure:'
  192241
```

```
Electrical Polarizability
   Optical Anisotropy
      Optical Anisotropy:  (Ref. 2)

                              - end of record -


      Display 21/CNGSSYKWIC/4

7,10-dimethyl-benz(c)acridine
7,10-dimethyl-benz(c) acridine
  'Graphic Structure:'
  192914
```

```
Electrical Polarizability
   Optical Anisotropy
      Optical Anisotropy:  (Ref. 4)


                            - end of record -
```

3D9. Nuclear Quadrupole Coupling Constant

The NQC field can currently be searched only for its comments, with the DP=NQCC search statement:

```
?s dp=nqcc
      S22     195  DP=NQCC  (Nuclear Quadrupole Coupling
                              Con.Com./refs.)
```

Searches for all halogen-containing compounds, for example, for which nuclear quadrupole data have been reported are possible:

```
?s (gn=a7 not gn=a6) and dp=nqcc
         354898  GN=A7
        1584038  GN=A6
            195  DP=NQCC  (Nuclear Quadrupole Coupling Con.
                            Com./refs.)
      S23      86  (GN=A7 NOT GN=A6) AND DP=NQCC

?d /bi k/15-25
       Display 23/BIK/15

1732392
chloro-trifluoro-methane
  German Chem. Name: Chlor-trifluor-methan
  Lawson No: 11
  Beilstein Cit: 3-01-00-00042; 4-01-00-00034
  Molecular Formula: CCLF3
  Molecular Weight: 104.46
  No. of Ref: 98
  Data Present:
   Data  Ref
   +Ref/Only UDF    Data Type
      26/       PR Preparative Data
      11/       CR Chemical Reactions
      40/24     PP Physical Properties
```

•   Molecular Formula: CClF3
Coupling Phenomena
   Nuclear Quadrupole Coupling Constant: Nuclei: 35Cl; (Ref.
   42 handbook) (ref. 43 handbook)

                              - end of record -
?
      Display 23/BIK/16

1732026
trideuterio-iodo-methane
  German Chem. Name: Trideuterio-jod-methan
  Lawson No: 9
  Beilstein Cit: 4-01-00-00091; 3-01-00-00098
  Molecular Formula: CD3I
  Molecular Weight: 144.96
  No. of Ref: 19
  Data Present:
   Data Ref
   +Ref/Only UDF   Data Type
      4/   PR Preparative Data
      2/   CR Chemical Reactions
     11/5   PP Physical Properties
  Molecular Formula: CD3I
Coupling Phenomena
   Nuclear Quadrupole Coupling Constant: Nuclei: 127I; (Ref.
   7 handbook
                              -more-

?d /cn k/25-29
      Display 23/CNK/25

2,2-dichloro-propane
  Molecular Formula: C3H6Cl2
Coupling Phenomena
   Nuclear Quadrupole Coupling Constant: Nuclei: 35Cl; (Ref.
   19 handbook)

                              - end of record -

?
      Display 23/CNK/26

tribromo-methane
  Molecular Formula: CHBr3
Coupling Phenomena

```
    Nuclear Quadrupole Coupling Constant:  Nuclei: 79Br; (Ref.
    34 handbook), (Ref. 35 handbook...

                            - end of record -


?
        Display 23/CNK/27

dichloro-fluoro-methane
   Molecular Formula:  CHCl2F
Coupling Phenomena
   Nuclear Quadrupole Coupling Constant:  Nuclei: 35Cl; (Ref.
   25 handbook)

                            - end of record -


?
        Display 23/CNK/28

chloro-difluoro-methane
   Molecular Formula:  CHClF2
Coupling Phenomena
   Nuclear Quadrupole Coupling Constant:  Nuclei: 35Cl; (Ref.
   28 handbook)

                            - end of record -


?
        Display 23/CNK/29

2-chloro-2-methyl-propane
   Molecular Formula:  C4H9Cl
Coupling Phenomena
   Nuclear Quadrupole Coupling Constant:  Nuclei: 35Cl; (Ref.
   26 handbook)

                            - end of record -
```

3D10. Moment of Inertia

All nonsymmetrical molecules possess a *moment of inertia*, which is a measure of the tendency of a rotating molecule to continue to rotate when the external force that caused the rotation is removed. The moment of inertia can be determined by microwave spectroscopy or calculated from the structure, and it determines the character of the microwave, or rotational, spectrum of the molecule. Moments of inertia have units of $g \cdot cm^2$ and are stored in the MI field of the Beilstein Database:

```
?s dp=mic
      S24      56  DP=MIC  (Moment Of Inertia Comments/refs.)

?d /cn sy kwic/5
      Display 24/CNSYKWIC/5

carbon selenide sulfide
  Synonym: Kohlenstoff-selenid-sulfid
Molecular Deformation & Potential
  Moment of Inertia: (Comment: (aus der Mikrowellen-
  Absorption ermittelt) von CS 76 Se, CS 77 Se...

                              - end of record -
```

It is possible, for example, to retrieve information leading to moment of inertia data on deuterium-containing compounds:

```
?s d1/ec and dp=mic
           3459  D1/EC
             56  DP=MIC  (Moment Of Inertia Comments/refs.)
      S25      7  D1/EC AND DP=MIC

?d /cn sy k /1-7
      Display 25/CNSYK/1

N-deuterio-formamide
  Molecular Formula:  CH2DNO
Molecular Deformation & Potential
  Moment of Inertia: (Comment: Traegheitsmomente der beiden
  Konformeren.) (Ref. 3 handbook)

                              - end of record -
?
      Display 25/CNSYK/2

O-Deuterio-methanol
  Molecular Formula:  CH3DO
Molecular Deformation & Potential
  Moment of Inertia: (Comment: des Molekuels.) (Ref. 3
  handbook)

                              - end of record -
?
      Display 25/CNSYK/3
```

```
O-Deuterio-(13 C)methanol
  Molecular Formula:  CH3DO
Molecular Deformation & Potential
  Moment of Inertia: (Comment: des Molekuels.) (Ref. 1
  handbook)

                               - end of record -

?
      Display 25/CNSYK/4

  Synonym: trans-1-Deuterio-propylenimin
  Molecular Formula:  C3H6DN
Molecular Deformation & Potential
  Moment of Inertia:  (Ref. 1)

                               - end of record -

?
      Display 25/CNSYK/5

  Synonym: (4-D)Thiazol
  Molecular Formula:  C3H2DNS
Molecular Deformation & Potential
  Moment of Inertia: (Ref. 2)

                               - end of record -

?
      Display 25/CNSYK/6

  Synonym: 5-Deutero-thiazole
  Molecular Formula:  C3H2DNS
Molecular Deformation & Potential
  Moment of Inertia: (Ref. 3)

                               - end of record -
```

and, if necessary, the associated references can also be listed:

```
?d 25/cn sy k rf
      Display 25/CNSYKRF/1

N-deuterio-formamide
  No. of Ref: 3
Refs.
   1, Kurland, Wilson, JCPSA6, J.Chem.Phys., 27(1957)585,587
```

```
   2, Miyazawa, NPKZAZ, Nippon Kagaku Zasshi, 76(1955)821, CA
      (1956)5407
   3, Kurland, Wilson, JCPSA6, J.Chem.Phys., 27(1957)585,586
  Molecular Formula:  CH2DNO
Molecular Deformation & Potential
  Moment of Inertia: (Comment: Traegheitsmomente der beiden
  Konformeren.) (Ref. 3 handbook)

                              - end of record -
```

3D11. Molecular Rotational Constant

The *molecular rotational constant* of a molecule is a quantity, measured in megahertz, which is proportional to the reciprocal of the moment of inertia. It is stored in the MRC field of the Beilstein Database, and this field can be searched for comments:

```
?s dp=mrcc
      S26    158  DP=MRCC   (Rotational Constants Comments
                              /refs.)

?d /cn sy k/2-3
      Display 26/CNSYK/2

trideuterio-nitro-methane
Molecular Deformation & Potential
  Rotational Constant: (Comment: und Potentialschwelle der
  inneren Rotation.) (Ref. 3 handbook)

                              - end of record -

?
      Display 26/CNSYK/3

dideuterio-diazo-methane
Molecular Deformation & Potential
  Rotational Constant: (Comment: aus dem Mikrowellen-
  spektrum.) (Ref. 3 handbook)

                              - end of display -
```

3D12. Energy Barrier of Conformation

Free rotation of bonds in a molecule, allowing it to flip from one conformation to another, is sometimes obstructed by parts of the structure. In such cases, the energy required to overcome this barrier has been reported and is stored in the EBC field. Records from this field can be retrieved with the DP=EBCC search statement:

```
?s dp=ebcc
        S27    647  DP=EBCC  (Energy Barriers Comments/refs.)

?d /cn sy k/51

?
        Display 27/CNSYK/51

   Synonym: N-Brom-3,3-dimethylazetidin
Molecular Deformation & Potential
    Energy Barrier: (Ref. 1)

                                  - end of record -
```

Information concerning the energy barriers to conformational change in piperidines can be retrieved with the search

```
?s piperidin?/cn and dp=ebcc
          12436  PIPERIDIN?/CN
            647  DP=EBCC  (Energy Barriers Comments/refs.)
      S28    15  PIPERIDIN?/CN AND DP=EBCC

?d /cn sy gs k /1-2
        Display 28/CNSYGSK/1

bis-(piperidine-1-thiocarbonyl)-disulfane
bis-( piperidine -1-thiocarbonyl)-disulfane
  'Graphic Structure:'
```

```
  Molecular Deformation & Potential
    Energy Barrier: (Ref. 10)

                              - end of record -

?
        Display 28/CNSYGSK/2
```

```
1-benzoyl-piperidin-4-one
1-benzoyl- piperidin -4-one
  'Graphic Structure:'
```

```
Molecular Deformation & Potential
   Energy Barrier: (Ref. 7)


                          - end of display -
```

3D13. Dissociation Energy

All chemical bonds have bond energies, and before a specific bond can be dissociated, its bond energy must be provided. *Bond dissociation energy* can be measured using a variety of direct and indirect methods. It is measured in joules per mole, and the parameter that is carried with it for display, but not search, is the *bond type*.

An example of a bond dissociation energy as it appears in the database is retrieved below using the "data present" (DP=) qualifier (see section 3A2):

```
?s dp=edis
       S29     32  DP=EDIS  (Dissociation Energy)

?d /k/1-2
       Display 29/K/1

Molecular Energy Parameters
   Dissociation Energy: 125600 J/mol; Bond Type: COO-OCO;
   (Ref. 4 handbook)

                              - end of record -
?
       Display 29/K/2

Molecular Energy Parameters
   Dissociation Energy: 125600 J/mol; Bond Type: COO-OCO;
   (Ref. 5 handbook)

                              - end of record -
```

Searches for specific values of bond dissociation energies can be carried out in the usual way:

```
?s edis=123510
      S30      1  EDIS=123510

?d /cn k
      Display 30/CNK/1

diacetyl peroxide
Molecular Energy Parameters
  Dissociation Energy:  123510  J/mol; Bond Type: RO-OR;
  (Ref. 24 handbook)

                        - end of display -
```

and ranges may be used for the energy value:

```
?s 370000<edis<400000
      S31      5  370000<EDIS<400000

?d /cn k/3
      Display 31/CNK/3

acetic acid propyl ester
Molecular Energy Parameters
  Dissociation Energy:  371790  J/mol; Bond Type: CO-OC3H7;
  (Ref. 27 handbook)

                        - end of display -
```

3D14. Ionization Potential

The energy required to remove one electron from a molecule is termed the *ionization potential* of the molecule and is commonly measured in electron volts (1 eV = 1.602×10^{-19} J). This property is important in a variety of techniques, such as mass spectrometry, and is measured either by electron impact ionization or by photoionization, among other methods. The ionization potential is stored in the IP field, and the *method of determination* can be specified in search statements without a qualifier:

```
?s 9.8<ip<10
      S32     13  9.8<IP<10
```

```
?d /k
      Display 32/K/1

Molecular Energy Parameters
  Ionization Potential:  9.97  eV; Method: Electron impact;
  (Ref. 28 handbook...

                              - end of display -
```

The ionization potential can be entered as a range of values:

```
?s ip=9:10
      S33        54  IP=9:10
```

and the method of determination can be specified, linked to the ionization potential value with an (S) operator to ensure that the two terms belong to the same experiment:

```
?s ip=9:10(s)photoionization
              54   IP=9 : IP=10
              44   PHOTOIONIZATION
      S34     13   IP=9:10(S)PHOTOIONIZATION

?s ip=9:10(s)(electron(w)impact)
              54   IP=9 : IP=10
             959   ELECTRON
             126   IMPACT
      S35     47   IP=9:10(S)(ELECTRON(W)IMPACT)

?d /cn sy k/1-2
      Display 35/CNSYK/1

hexan-2-one
Molecular Energy Parameters
  Ionization Potential:  9.58  eV; Method: Electron impact;
  (Ref. 48 handbook)

                              - end of record -

?
      Display 35/CNSYK/2
```

```
hexa-1,5-diene
Molecular Energy Parameters
   Ionization Potential:  9.51  eV; Method: Electron impact;
   (Ref. 26 handbook)

                                   - end of display -
```

3D15. Proton Affinity and Electron Affinity

All molecules have an affinity both for protons and for electrons, and the degree of this affinity determines the behavior of the molecule towards protons or electrons. Affinity data are stored in the RTAFF field of the Beilstein Database and may be retrieved with the DP=RTAFF statement:

```
?s dp=rtaff
       S36    123  DP=RTAFF  (Affinity)

?d /k/1-3
       Display 36/K/1

Molecular Energy Parameters
   Affinity
       Electron affinity (Ref. 1)

                             - end of record -
?
       Display 36/K/2

Molecular Energy Parameters
   Affinity
       Proton affinity (Ref. 1)

                             - end of record -
```

All proton affinities can be retrieved by using the word "proton":

```
?s dp=rtaff and proton
           123  DP=RTAFF  (Affinity)
            69  PROTON
       S37   53  DP=RTAFF AND PROTON
```

and all electron affinities can be located similarly:

```
?s dp=rtaff and electron
          123  DP=RTAFF  (Affinity)
          959  ELECTRON
    S38    70  DP=RTAFF AND ELECTRON
```

There are no affinities which are not associated with either protons or electrons:

```
?s dp=rtaff not (proton or electron)
          123  DP=RTAFF  (Affinity)
           69  PROTON
          959  ELECTRON
    S39     0  DP=RTAFF NOT (PROTON OR ELECTRON)
```

3E. MULTICOMPONENT SYSTEMS

Systems such as solutions and liquid mixtures, which have more than one component, are regarded in the Beilstein Database as *multicomponent Systems*, and a number of special fields are reserved for such systems. In each case a property, such as

TABLE XI

PROPERTY	SEARCH FIELD	DISPLAY FIELDS		
Solution behavior	SOLNB	MC	SL	
Solubility	SL	MC	SL	
Solubility product	SLP	MC	SL	
Liquid–liquid systems	LL	MC	LL	
Liquid–solid systems	LS	MC	LS	
Liquid–vapor systems	LV	MC	LV	
Transport phenomena	MCTP	MC		
Azeotropes	AZ	MC	LV	AZ
Other mechanical properties	MCOM	MC		
Energy data	MCEDATA	MC	MC	
Critical micellar concn.	CMCC			
Boundary surface phenomena	MCBS	MC		
Adsorption	ADSP	MC		
Association	ASSN	MC		

azeotrope formation, is associated with one or more parameters, such as concentration. It is possible to search for both properties and parameters; the two can be linked by means of the (S) operator. Thus a search statement like

s dp=az(s)water (s)temp=50:60

will retrieve compounds which, with water, form an azeotrope, b.p. 50–60°C.
     Fields available for multicomponent systems are listed in Table XI.

3E1. Solution Behavior

Many compounds have a variety of solubility properties which are gathered together in the Beilstein Database under the general heading of solution behavior and stored in the SOLNB field. This field contains uncontrolled vocabulary, and it can be searched for the presence of data:

```
?s dp=solnb
     S1    1430   DP=SOLNB  (Solution Behavior)

?d /k/400-402
        Display 1/K/400

Solution Behavior
    Miscibility; Partner: ethanol; Temp: 20 - 50 C (Ref. 28
    handbook...

                              - end of record -

?
        Display 1/K/401

Solution Behavior
    Miscibility; Partner: ethanol; Temp: 20 - 30 C (Ref. 76
    handbook...

                              - end of record -

?
        Display 1/K/402

Solution Behavior
    Dissolving  capacity; Partner: nitrocellulose (Ref. 6
    handbook...

                              - end of record -
```

The body of the SOLNB entry can be searched for specific terms, such as "methanol" or "benzene". Such terms are unqualified, and if they are linked to the DP=SOLNB term with an AND operator, false drops will result:

```
?s dp=solnb and ethanol
          1430   DP=SOLNB  (Solution Behavior)
        234177   ETHANOL
     S2    722   DP=SOLNB AND ETHANOL

?d /k/10
        Display 2/K/10

Preparative Data
   ...Preparation
       Starting Material: glycerol, ketene
       Reagent: acetone, ethanol
       By-product: monoacetyne, diacetyne (Ref. 12 handbook...
Crystals
   ...Melting Point: 3.2 C; Solvent: ethanol; (Ref. 29
       handbook...
Solution Behavior
       Dissolving capacity; Partner: air; Temp: 20 C (Comment:
       Loest 9.4 Vol.-prozent Luft.) (Ref. 64...
Chemical Reactions
   ...Chemical Reaction:
       Partner: ethanol , HCl
       Further Conditions: Geschwindigkeit der Alkoholyse
       (Ref. 97 handbook
   ...Chemical Reaction:
       Partner: ethanol, KOH
       Further Conditions: Alkoholyse (Ref. 98 handbook...

                            - end of display -
```

This record has "ethanol" in four places, but not in the section on solution behavior. It therefore constitutes a false drop. If the (S) operator is used in place of the AND operator, such false drops do not materialize:

```
?s dp=solnb(s)ethanol
          1430   DP=SOLNB  (Solution Behavior)
        234177   ETHANOL
     S3     56   DP=SOLNB(S)ETHANOL

?d 3/gs bi k/6
        Display 3/GSBIK/6
```

```
1802063
1,2,3-tris-nitryloxy-propane
  German Chem. Name: 1,2,3-Tris-nitryloxy-propan
  Lawson No: 636
  Beilstein Cit: 3-01-00-02328; 1-01-00-00272; 4-01-00-02762;
  0-01-00-00516; 2-01-00-00591
  Molecular Formula: C3H5N3O9
  Molecular Weight: 227.09
  No. of Ref: 122
Graphic Structure:
'1802063'
```

```
?

       Display 3/GSBIK/6


  Data Present:
   Data  Ref
   +Ref/Only UDF    Data Type
     10/       PR Preparative Data
     47/       CR Chemical Reactions
    143/13    PP Physical Properties
Electronic Spectra
  UV/VIS Spectrum
  ...Solvent:  ethanol ; (Ref. 61 handbook...
Solution Behavior
  ...Solubility:  In Solution or in Pure Solvent  Temp: 15 C;
     Solvent: ethanol ; (Comment: Solubility :25.0 Gew.-
     percent.) (Ref. 65 handbook
  ...Solubility: In Solution or in Pure Solvent  Temp: 40 C;
     Solvent:  ethanol ; (Comment: Solubility :41.4 Gew.-
     percent.) (Ref. 65 handbook
  ...Solubility: In Solution or in Pure Solvent  Temp: 0 C;
     Solvent:  ethanol ; (Comment:  100 g solvent dissolves.
     37.5 g Substance.) (Ref. 66 handbook
  ...Solubility: In Solution or in Pure Solvent  Temp: 20 C;
     Solvent: ethanol; (Comment: 100 g solvent
```

                              -more-

```
?

      Display 3/GSBIK/6

      dissolves. 54 g Substance.) (Ref. 66 handbook
   ...Solubility: In Solution or in Pure Solvent  Temp: 20 C;
      Solvent: ethanol; (Comment: 100 g solvent dissolves.
      40 g Substance.) (Ref. 66 handbook
   ...Solubility: 250 g/l; In Solution or in Pure Solvent
      Solvent: ethanol; (Ref. 12 handbook...
   ...Dissolving  capacity; Partner:  ethanol ; Temp: 15 - 40
      C (Comment: 3.2 - 9.7 Gew.-prozent.) (Ref.65
      handbook...

                              - end of display -
```

3E2. Solubility

Solubility data from the literature are carried in the SL field of the Beilstein Database.
An availability search in this field, using the DP= qualifier (see section 3A2), produces
all the records which include solubility information:

```
?s dp=sl
       S4    2009  DP=SL  (Solubility)

?d /sl/109
       Display 4/SL/109

Solution Behavior
   Solubility: 0.2 g/l; In Solution or in Pure Solvent  Temp:
   25 C; Solvent: H2O; (Ref. 2 handbook)
   Solubility: 64.8 g/l; In Solution or in Pure Solvent Temp:
   25 C; Solvent: ethanol; (Ref. 2 handbook)

                              - end of display -
```

The two parameters which accompany the solubility data are the *solvent* and the
temperature at which the solubility was measured (TEMP=). These may be used in
searches either independently:

```
?s dp=sl(s)acetone
          2009  DP=SL  (Solubility)
         47143  ACETONE
     S5     45  DP=SL(S)ACETONE
```

```
?d /k
        Display 5/K/1

Solution Behavior
    Solubility: 4 g/l; In Solution or in Pure Solvent  Temp:
    25 C; Solvent: acetone ; (Ref. 3 handbook)

                              - end of display -
```

or together:

```
?s dp=sl(s)ethanol(s)temp=15:20
            2009   DP=SL  (Solubility)
          234177   ETHANOL
          146986   TEMP=15 : TEMP=20
    S6       105   DP=SL(S)ETHANOL(S)TEMP=15:20

?d /cn sy mc/1-3
        Display 6/CNSYMC/1

allophanic acid tert-pentyl ester
Solution Behavior
    Solubility: 11.25 g/l; In Solution or in Pure Solvent
    Temp: 15.5 C; Solvent: ethanol; (Ref. 4 handbook)
    Solubility: 2.89 g/l; In Solution or in Pure Solvent
    Temp: 16 C; Solvent: diethyl ether; (Ref. 4 handbook)
    Refs.
        4, Behal, BSCFAS, Bull.Soc.Chim.Fr., (4)25 (1919),478

                              - end of record -

?
        Display 6/CNSYMC/2

allophanic acid tert-butyl ester
Solution Behavior
    Solubility: 7.23 g/l; In Solution or in Pure Solvent
    Temp: 15 C; Solvent: ethanol; (Ref. 2 handbook)
    Solubility: 2.9 g/l; In Solution or in Pure Solvent  Temp:
    15 C; Solvent: diethyl ether; (Ref. 2 handbook)
    Solubility:  In Solution or in Pure Solvent  Solvent:
    ethanol; (Comment: 1 part(s) of substance.dissolves in:10
    parts of solvent.in boiling solvent.) (Ref. 2 handbook)
    Refs.
        2, Behal, BSCFAS, Bull.Soc.Chim.Fr., (4)25 (1919),478

                              - end of record -
```

A typical use of this field is illustrated in the search shown below, in which information concerning the solubility of glucose in ethanol is retrieved:

```
?s glucose/cn and dp=sl(s)ethanol
            607   GLUCOSE/CN
           2009   DP=SL  (Solubility)
         234177   ETHANOL
            257   DP=SL(S)ETHANOL
    S7          1   GLUCOSE/CN AND DP=SL(S)ETHANOL

?d /cn k
        Display 7/CNK/1

D-Glucose
D- Glucose
Solution Behavior
    Solubility : 20 g/l; In Solution or in Pure Solvent  Temp:
    20 C; Solvent: aq. ethanol (80percent); (Comment: in
    reinem Loesungsmittel: alpha -Glucose.) (Ref. 97
    handbook...

                          - end of display -
```

3E3. Solubility Product

For solutes that ionize in solution, the *solubility product* is defined as $[A]^a[B]^b$, where A and B are the ions and *a* and *b* are their charges, respectively. Thus for AgCl or NaCl, the solubility product is the square of the molar concentration. Since most organic compounds fail to exhibit ionic behavior, solubility product is not a common property in the Beilstein Database. Where it is reported, however, it is maintained in the SLP field; the *solvent* is also identified, and the *temperature* of the measurement is in the TEMP= field:

```
?s dp=slp
    S8          1   DP=SLP   (Solubility Product)

?d /cn sy k
        Display 8/CNSYK/1

4 alpha,9-epoxy-3 beta-(2-methyl-cis-crotonoyloxy)-5 beta-
cevane-4 beta,12,14,16 beta,17,20-hexaol
```

```
Solution Behavior
   Solubility Product: 0.6E-7; Temp: 15 C; Solvent: H2O;
   (Ref. 45 handbook)

                              - end of display -
```

The other parameters that are relevant to solubility product are solvent and temperature (TEMP=). They can be searched in conjunction with a solubility product search in the usual way:

```
?s dp=slp(s)temp=15(s)h2o
               1   DP=SLP  (Solubility Product)
            6812   TEMP=15
           67329   H2O
      S9       1   DP=SLP(S)TEMP=15(S)H2O
```

3E4. Liquid–Liquid Systems

Information concerning the properties of liquid–liquid systems is stored in the LL field. This field is not searchable, but data present in the field can be detected and displayed:

```
?s dp=LL
      S10    2295   DP=LL   (Liquid/liquid Systems)

?d 10/cn sy mc/950-952
        Display 10/CNSYMC/950

   Synonym: Rodiasine
Solution Behavior
    Solubility: (Ref. 1)
    Refs. 1, Hearst, JOCEAH, J.Org.Chem., 29(1964)466
Liquid/Liquid System
    Distribution between solvent 1 + 2 (Ref. 1)
    Refs. 1, Hearst, JOCEAH, J.Org.Chem., 29(1964)466

                              - end of record -

?

        Display 10/CNSYMC/951
```

```
Liquid/Liquid System
   Distribution between solvent 1 + 2 (Ref. 1)
   Refs. 1, Takai, et al, JMCMAR, J.Med.Chem., 22(1979)1380,
   1381,1382,1383

                                - end of record -
?
      Display 10/CNSYMC/952

Liquid/Liquid System
   Distribution between solvent 1 + 2 (Ref. 1)
   Refs. 1, Akerblom, JMCMAR, J.Med.Chem., 17(1974)609,610,614

                                - end of display -
```

3E5. Liquid–Solid Systems

Information on liquid–solid systems is stored in the LS field. This field is not search-able, but data present in it can be retrieved and displayed:

```
?s dp=ls
         S11    1534   DP=LS  (Liquid/solid Systems)

?d /k/500-504
         Display 11/K/500

Liquid/Solid System
    Liquid /solid phase diagram; Partner: dotriacontane (Ref.
    9 handbook)

                                - end of record -
```

This field is useful when it is necessary to find the few relevant records from a larger set derived by a chemical name search. In the example shown below, the search retrieves all liquid–solid system records in which dotriacontane is mentioned:

```
?s dp=ls(s)dotriacontane
           1534   DP=LS  (Liquid/solid Systems)
             44   DOTRIACONTANE
     S12      3   DP=LS(S)DOTRIACONTANE
```

```
?d /bi k/1
        Display 12/BIK/1

1730718
propane
  German Chem. Name: Propan
  Lawson No: 27
  Beilstein Cit: 3-01-00-00204; 4-01-00-00176; 2-01-00-00071;
  0-01-00-00103; 1-01-00-00033
  Molecular Formula: C3H8
  Molecular Weight: 44.1
  No. of Ref: 736
  Data Present:
   Data  Ref
   +Ref/Only UDF    Data Type
     98/        PR Preparative Data
     260/       CR Chemical Reactions
     373/105  PP Physical Properties
Liquid/Solid System
    Solidification  points of mixtures; Partner: dotriacontane
    (Ref. 494 handbook...

                                - end of record -
```

3E6. Liquid–Vapor Systems

All information on liquid–vapor systems that has been published is collected in the LV field. Data in this field can be retrieved with the DP= qualifier:

```
?s dp=lv
        S13     517  DP=LV  (Liquid/vapor Systems)

?d /cn/200-202
        Display 13/CN/200

hexafluoro-propene

                        - end of record -
```

The LV field can be used to find which records from a set contain data on liquid–vapor systems:

```
?s mf=c4h10o and dp=lv
            10   MF=C4H10O
           517   DP=LV  (Liquid/vapor Systems)
     S14     6   MF=C4H10O AND DP=LV

?d /bi k/6
        Display 14/BIK/6

773649
butan-2-ol
  German Chem. Name: Butan-2-ol
  Lawson No: 317
  Beilstein Cit: 2-01-00-00400; 0-01-00-00371; 1-01-00-00188
  Molecular Formula: C4H10O
  Molecular Weight: 74.12
  No. of Ref: 118
  Data Present:
   Data  Ref
   +Ref/Only UDF    Data Type
      36/        PR Preparative Data
      45/        CR Chemical Reactions
      67/4       PP Physical Properties
  Molecular Formula:  C4H10O
Liquid/Vapor System
    Vapour pressure diagram for the mixture; Partner:
    nitrobenzene; Temp: 30 - 49.7 C (Ref. 52 handbook...

                            - end of display -
```

Words within the LV field can be used in search statements. In the example below, a search is made for all isomers of butanol for which data on their behavior with water in a liquid–vapor system are recorded:

```
?s mf=c4h10o and dp=lv(s)water
            10   MF=C4H10O
           517   DP=LV  (Liquid/vapor Systems)
         31566   WATER
            63   DP=LV(S)WATER
     S15     2   MF=C4H10O AND DP=LV(S)WATER

?d /bi k/1-2
        Display 15/BIK/1
```

```
1730878
2-methyl-propan-1-ol
  German Chem. Name: 2-Methyl-propan-1-ol
  Lawson No: 317
  Beilstein Cit: 3-01-00-01550; 2-01-00-00405; 4-01-00-01588;
  0-01-00-00373; 1-01-00-00189
  Molecular Formula: C4H100
  Molecular Weight: 74.12
  No. of Ref: 608
  Data Present:
   Data  Ref
   +Ref/Only UDF    Data Type
     56/        PR Preparative Data
    174/        CR Chemical Reactions
    465/87      PP Physical Properties
     12/        DR Characterization Derivative
  Molecular Formula:  C4H100
Liquid/Vapor System
    Liquid/vapour phase diagram; Partner:  water (Ref. 483
    handbook), (Ref. 484 handbook),  (Ref. 485 handbook...

                            - end of record -
```

3E7. Transport Phenomena

Transport phenomena exhibited by multicomponent systems—typically binary systems
—are recorded in the MCTP field. This field can be retrieved with the DP= qualifier, or
words within the field can be used as search terms:

```
?s dp=mctp
      S16      772   DP=MCTP  (Transport Phenomena)

?d /cn k/34-35
       Display 16/CNK/34

1,6-bis-stearoyloxy-hexane
Transport Phenomena
    Viscosity; Solvent: benzene (Ref. 1 handbook),  (Ref. 2
    handbook...

                            - end of record -
?
       Display 16/CNK/35
```

```
decanedioic acid dihexadecyl ester
Transport Phenomena
    Viscosity ; Partner: CCl4; Temp: 20 C (Ref. 3 handbook),
    (Ref. 4 handbook...

                              - end of record -
```

In the example shown below, the search for a tridecyl ester of a group III element (boron, aluminum, etc.) produced numerous compounds, but only one of them carried any transport property data on the mixtures formed by the compound, in this case with tetrachloromethane:

```
?s (tridecyl(w)ester)/cn and gn=a3 and dp=mctp
             241   TRIDECYL/CN
           68684   ESTER/CN
              24   TRIDECYL/CN(W)ESTER/CN
            4148   GN=A3
             772   DP=MCTP   (Transport Phenomena)
     S17       1   (TRIDECYL(W)ESTER)/CN AND GN=A3 AND DP=MCTP


?d /bi gs k
        Display 17/BIGSK/1


1715121
boric acid tridecyl ester
   German Chem. Name: Borsaeure-tridecylester
   Lawson No: 362
   Beilstein Cit: 4-01-00-01821; 3-01-00-01761
   Molecular Formula: C30H63BO3
   Molecular Weight: 482.64
   Synonym: Borsaeure-trisdecylester
   No. of Ref: 5
   'Graphic Structure:'
   1715121
```

```
   Data Present:
    Data  Ref
    +Ref/Only UDF     Data Type
       13/      PP Physical Properties
boric acid tridecyl ester
  Molecular Formula:  C30H63BO3
Transport Phenomena
     Viscosity; Partner: tetrachloromethane (Ref. 5
handbook...


                      - end of display -
```

3E8. Azeotropes

Many chemicals, when combined in a liquid mixture, form *azeotropes*, or constant boiling mixtures. These are mixtures which, when boiled, form vapors with the same concentration; the components cannot be separated by distillation. Any material in the Beilstein Database which is known to form azeotropes is so identified. The name of the other component of the azeotrope will be found in the AZ field. The concentration of the azeotropic mixture is also in the AZ field, as are its boiling point and the operative pressure:

```
?s dp=az
      S18     929  DP=AZ  (Azeotrope Components)

?d /k/70
      Display 18/K/70

Liquid/Vapor System
   Azeotropes:
      Components: water; Temp: 98.8 C; Pressure: 760 Torr;
      (Ref. 28 handbook),  (Ref. 29...

                      - end of record -
```

If one is seeking azeotropes involving water, then the term "water" may be added to the search statement, linked with an (S) proximity operator:

```
?s dp=az(s)water
            929  DP=AZ  (Azeotrope Components)
          31566  WATER
      S19    413  DP=AZ(S)WATER
```

```
?d /k/100
        Display 19/K/100

Liquid/Vapor System
   Azeotropes:
       Components: water; (Ref. 13 handbook)

                            - end of record -
```

The temperature and pressure at which an azeotrope boils can also be specified, as in the examples below:

```
?s dp=az(s)water(s)temp=78.7
              929   DP=AZ   (Azeotrope Components)
            31566   WATER
                7   TEMP=78.7
       S20      1   DP=AZ(S)WATER(S)TEMP=78.7

?s dp=az(s)water(s)temp=78.7(s)pres=760
              929   DP=AZ   (Azeotrope Components)
            31566   WATER
                7   TEMP=78.7
              575   PRES=760
       S21      1   DP=AZ(S)WATER(S)TEMP=78.7(S)PRES=760

?d /bi k
        Display 21/BIK/1

1718880
1,2-dichloro-propane
  German Chem. Name: 1,2-Dichlor-propan
  Lawson No: 28
  Beilstein Cit: 3-01-00-00225; 0-01-00-00105; 2-01-00-00073
  Molecular Formula: C3H6Cl2
  Molecular Weight: 112.99
  General Comment: stereoisomeres of unknown configuration.
       stereoisomeres of unknown configuration.
       stereoisomeres of unknown configuration.
  No. of Ref: 80
  Data Present:
   Data  Ref
   +Ref/Only UDF    Data Type
      25/      PR Preparative Data
      21/      CR Chemical Reactions
      54/4     PP Physical Properties
Liquid/Vapor System

                        -more-
```

```
?
        Display 21/BIK/1

   Azeotropes:
       Components: water; Temp: 78.7 C; Pressure: 760 Torr;
       (Comment: 1.2-dichloro-propane) (Ref. 58 handbook),
       (Ref. 59 handbook...

                            - end of display -
```

3E9. Other Mechanical Properties

Other mechanical properties of multicomponent systems include a variety of phenomena such as compressibility, volume change on mixing, and *PVT* relationships. These are all collected together in the MCOM field. This field can be searched using the DP= qualifier:

```
?s dp=mcom
        S22    251   DP=MCOM  (Other Mechanical Properties)

?d /k/50
        Display 22/K/50

Other Mechanical Properties
    Adiabatic compressibility; Solvent: H2O; Temp: 30 C (Ref.
    117 handbook)

                            - end of record -
```

Terms within the MCOM field can be used in searches:

```
?s dp=mcom(s)compressibility(s)water
          251   DP=MCOM  (Other Mechanical Properties)
          326   COMPRESSIBILITY
        31566   WATER
    S23     3   DP=MCOM(S)COMPRESSIBILITY(S)WATER

?d /bi k
        Display 23/BIK/1

1209246
formic acid
  German Chem. Name: Ameisensaeure
  Lawson No: 1145
```

```
  Beilstein Cit: 3-02-00-00003; 4-02-00-00003; 2-02-00-00003;
  1-02-00-00007; 0-02-00-00008
  Molecular Formula: CH2O2
  Molecular Weight: 46.03
  No. of Ref: 813
  Data Present:
   Data  Ref
   +Ref/Only UDF    Data Type
     59/       PR Preparative Data
     229/      CR Chemical Reactions
     563/114   PP Physical Properties
Other Mechanical Properties
    Adiabatic compressibility; Partner: water; Temp: 30 - 34
    C (Ref. 650 handbook)


                        - end of display -
```

All data dealing with the mixing of compounds with water can be retrieved by the search below:

```
?s dp=mcom(s)mixing(s)(water or h2o)
            251   DP=MCOM   (Other Mechanical Properties)
            255   MIXING
          31566   WATER
          67329   H2O
    S24      27   DP=MCOM(S)MIXING(S)(WATER OR H2O)

?d /cn k /1-2
      Display 24/CNK/1

pentane-1,2-diol
Other Mechanical Properties
    Volume change on mixing; Partner: water; Temp: 20 C (Ref.
    2 handbook...

                          - end of record -
?
      Display 24/CNK/2

D-glucitol
Other Mechanical Properties
    Volume change on mixing; Partner: borax; Solvent: H2O
    (Ref. 133 handbook...

                          - end of record -
```

```
?
      Display 29/BIK/2

1731048
tribromo-methane
Energy Data
   Heat capacity of mixtures; Partner: acetone (Ref. 241
   handbook...

                            - end of display -
```

3E11. Micelles

Some compounds, such as those with a long hydrocarbon chain linked to a hydrophilic
end group (e.g. sodium dodecylsulfate) may, under proper conditions, form large scale
aggregates called *micelles*. The concentration in solution above which this is possible
is called the *critical micellar concentration*, and this parameter is carried in the
Beilstein Database in the field CMC. The *solvent* in question is identified in the field, as
is the operative *temperature*, in degrees Celsius.
     The CMC field can be searched with the dp=cmcc statement:

```
?s dp=cmcc
      S30      40  DP=CMCC  (Critical Micelle Concentration

?d /cn k/33
      Display 30/CNK/33

butyric acid
Solution Behavior
   Critical Micelle Concentration: Temp: 0 C; Solvent: H2O;
   (Ref. 188 handbook...

                            - end of record -
```

3E12. Boundary Surface Phenomena

Various boundary surface phenomena, such as pressure–surface isotherms or interfa-
cial tension, are recorded in the literature and stored in the MCBS field. This field can be
searched with the dp= qualifier:

```
?s dp=mcbs
      S31     940  DP=MCBS  (Boundary Surface Phenomena)
```

```
?d /k/45
      Display 31/K/45

Boundary Surface Phenomena
    Pressure-surface isotherm (Comment: einer monomolekularen
    Schicht an der Grenzflaeche Petrolaether/wss. Salzsaeure
    sowie an der...

                              - end of record -
```

and terms from within the MCBS field can be used in searches:

```
?s dp=mcbs(s)"pressure-surface isotherm"
            940   DP=MCBS  (Boundary Surface Phenomena)
            302   PRESSURE-SURFACE ISOTHERM
     S32    267   DP=MCBS(S)"PRESSURE-SURFACE ISOTHERM"

?d /cn k/30
      Display 32/CNK/30

methyl-octadecyl-malonic acid
Boundary Surface Phenomena
    Pressure-surface isotherm; Partner: HCl (0.01 n)
    (Comment: monomolekularer Filme.) (Ref. 3 handbook...

                              - end of display -

?s dp=mcbs(s)"pressure-surface isotherm"(s)partner=water
            940   DP=MCBS  (Boundary Surface Phenomena)
            302   PRESSURE-SURFACE ISOTHERM
           8242   PARTNER=WATER
     S33     85   DP=MCBS(S)"PRESSURE-SURFACE ISOTHERM"(S)
                  PARTNER=WATER

?d /cn k/30-35
      Display 33/CNK/30

4-methyl-octadecanoic acid
Boundary Surface Phenomena
    Pressure-surface isotherm; Partner: water; Temp: 10 C;
    Comment: einer monomolekularen Schicht.) (Ref. 9
    handbook)

                              - end of record -
```

3E13. Adsorption

Data concerning the phenomenon of *adsorption*, in which gases or liquids interact with the surface of solids, are stored in the **ADSP** field. This field can be searched with the **dp=** qualifier:

```
?s dp=adsp
        S34       697  DP=ADSP  (Adsorption)

?d 34/k/56
        Display 34/K/56

Adsorption
    adsorption; Partner: activated MgO; Solvent: benzene; Temp:
    25 C (Comment: Geschwindigkeit.) (Ref. 186 handbook...

                            - end of record -
```

but searches are often more useful when there is additional specification such as the *partner* —the solid phase involved:

```
?s dp=adsp(s)partner="activated mgo"
            697  DP=ADSP  (Adsorption)
              1  PARTNER=ACTIVATED MGO
     S35      1  DP=ADSP(S)PARTNER="ACTIVATED MGO"
```

or the *solvent*:

```
?s dp=adsp(s)solvent=benzene
            697  DP=ADSP  (Adsorption)
          52086  SOLVENT=BENZENE
     S36      9  DP=ADSP(S)SOLVENT=BENZENE
```

This field can be used to find, for example, all compounds for which there are data concerning their adsorption by charcoal:

```
?s dp=adsp(s)partner=charcoal
            697  DP=ADSP  (Adsorption)
            245  PARTNER=CHARCOAL
     S37    132  DP=ADSP(S)PARTNER=CHARCOAL

?d /cn k/50-51
        Display 37/CNK/50
```

```
L-glutamic acid
Adsorption
    adsorption; Partner: charcoal; Solvent: H2O (Ref. 133
    handbook), (Ref. 134 handbook...

                                - end of record -
?
      Display 37/CNK/51

DL-aspartic acid
Adsorption
    adsorption; Partner: charcoal; Solvent: H2O (Ref. 32
    handbook), (Ref. 33 handbook

                         - end of record -
```

3E14. Association

Chemical compounds often interact with one another in an ill-defined process which, for want of a better name, is termed *association*. Data concerning the association of the title compound with other species are carried in the ASSN field. This field can be searched with the dp= qualifier:

```
?s dp=assn
     S38    1756  DP=ASSN  (Association)

?d /k/300
      Display 38/K/300

Association
    Stability constant of the complex with ... (Ref. 9)
                           - end of record -
```

The ASSN field frequently contains information as to the partner species, and this can be used in searches such as the one below, in which six examples are retrieved of the association of disulfides with iodine:

```
?s sulfide/cn and dp=assn(s)partner=iodine
           1654  SULFIDE/CN
           1756  DP=ASSN  (Association)
           1172  PARTNER=IODINE
             37  DP=ASSN(S)PARTNER=IODINE
     S39      6  SULFIDE/CN AND DP=ASSN(S)PARTNER=IODINE
```

```
?d /gs k/4-5
        Display 39/GSK/4


diisopentyl sulfide
  'Graphic Structure:'
  1698015
```

```
Association
    Association with compound; Partner: iodine; Solvent:
    2,2,4-trimethyl-pentane (Ref. 36 handbook),  (Ref. 37
    handbook...


                                - end of record -
?
        Display 39/GSK/5


ethyl-methyl sulfide
  'Graphic Structure:'
  1696871
```

```
Association
    Association with compound; Partner: iodine; Solvent: CCl4
    (Ref. 46 handbook)

                            - end of display -
```

3F. PHYSIOLOGICAL DATA

A good deal of physiological data on specific chemicals can be found in the Beilstein Database. This includes biological and toxicological data as well as data pertaining to the use and ecological effects of different chemicals. These data are gathered together under the heading of general information and are listed in Table XII.

Each of the various fields can be displayed or searched.

TABLE XII

PROPERTY	SEARCH FIELD	DISPLAY FIELDS		
Biological function	/BF	19	PB	BF
Ecological data	/ED	19	PB	ED
Toxicity	/TX	19	PB	TX
Use	/US	19	PB	US

3F1. Biological Function

When a publication contains information concerning *biological properties* of a compound, the field for biological function (BF) in the database is tagged. A very terse summary of the published biological activity is included, usually as a short phrase, such as "antibakteriell aktiv". This information can be used for searching, either with the dp= qualifier or with specific search terms:

```
?s dp=bf
      S1    4290  DP=BF  (Biological Function)

?d /k/474
        Display 1/K/474

Physiological Data
   Biological Function
      Fungizid  (Ref. 1)

                          - end of record -
```

Searches can be performed that will isolate families of compounds, together with their reported biological function. In the first example, five *N*-oxides of quinoline are retrieved and their biological function is displayed:

```
?s (quinoline and 1 and oxide)/cn and dp=bf
        15159  QUINOLINE/CN
       186709  1/CN
         4471  OXIDE/CN
```

```
                   4290   DP=BF  (Biological Function)
         S2          5   (QUINOLINE AND 1 AND OXIDE)/CN AND DP=BF



?d /cn bf/2-3
         Display 2/CNBF/2


4-nitro-quinoline-1-oxide
Physiological Data
   Biological Function
      fungizide Wirksamkeit (Ref. 7)
      antibakterielle Aktiv. (Ref. 43)
      Refs.
          7, Fukui, et al, BCSJA8, Bull.Chem.Soc.Jpn., 33(1960)122
          43, Coutts, et al, CJCHAG, Can.J.Chem., 48(1970)2393,2395


                              - end of record -

?
         Display 2/CNBF/3

5-nitro-quinoline-1-oxide
Physiological Data
   Biological Function
      antibakt. Aktiv. (Ref. 4)
      Refs.
          4, ??Coutts, et al, CJCHAG, Can.J.Chem., 48(1970)2393,2395


                              - end of display -
```

In the example shown below, the biological function of some barbiturates is retrieved:

```
?s (barbitur?/cn) and dp=bf
          2959   BARBITUR?/CN
          4290   DP=BF  (Biological Function)
     S3      6   (BARBITUR?/CN) AND DP=BF

?d /cn gs bf/1-2
         Display 3/CNGSBF/1
```

5-cyclohex-1-enyl-1,5-dimethyl-barbituric acid
  'Graphic Structure:'
  253102

Physiological Data
  Biological Function
      Einwirkung von Kaninchenleberschnitten -)
          3,5-Dimethyl-5-(3$a-hydroxy-cyclohexen-(1)-yl)-
          barbitursaeure,
          3,5-Dimethyl-5-(3-oxo-cyclohexen-(1)-yl)-barbitursaeure,
          5-Methyl-5-(3-oxo-cyclohexen-(1)-yl)-barbitursaeure (Ref.
          71)

      Verabreichung an Kaninchen: im Harn der Tiere:
          3,5-Dimethyl-5-(3$a-hydroxy-cyclohexen-(1)-yl)-
          barbitursaeure,
          3,5-Dimethyl-5-(3-oxo-cyclohexen-(1)-yl)-barbitursaeure,
          5-Methyl-5-(3-oxo-cyclohexen-(1)-yl)-barbitursaeure (Ref.
          71)

      Einwirkung von Enzympraeparaten aus Kaninchenleber -)
          5-(3-Hydroxy-cyclohexen-(1)-yl)-3,5-dimethyl-
          barbitursaeure,
          5-(3-Oxo-cyclohexen-(1)-yl)-3,5-dimethyl-barbitursaeure
          (Ref. 8)

      Einwirkung von Rattenleberschnitten (auch nach Vorbehandlung
      der Tiere m. Phenobarbital) -)
          3,5-Dimethyl-5-(3-hydroxy-cyclohexen-(1)-yl)-
          barbitursaeure,
          3,5-Dimethyl-5-(3-oxo-cyclohexen-(1)-yl)-barbitursaeure
          (Ref. 72)

                              -more-
?

      Display 3/CNGSBF/1

      Refs.
          8, Toki, et al, CPBTAL, Chem.Pharm.Bull., 10(1962)708,709
          71, Kuroiwa, CPBTAL, Chem.Pharm.Bull., 11(1963)160,162
          72, Okui, Kuroiwa, CPBTAL, Chem.Pharm.Bull.,
              11(1963)163,165

                          - end of record -

3F2. Ecological Data

When ecological data appear in a publication, an appropriate summary is entered into the ED field of the Beilstein Database. This field generally contains terse phrases describing the nature of the published report and can be searched with such words or phrases:

```
?s dp=ed
      S4        6  DP=ED  (Ecological Data)

?d /k/2
      Display 4/K/2

Physiological Data
   Ecological Data
      Obstbauinsektizid (Tab.1) (Ref. 2)

                              - end of record -
```

Words or phrases from the ED field can be used in searches such as the one below for "Blattinsektizid" (leaf insecticides):

```
?s blattinsektizid/ed
      S5        1  BLATTINSEKTIZID/ED

?d /bi ed
      Display 5/BIED/1

1087535
   Lawson No: 32101, 692, 310, 298
   Beilstein Cit: 5-27
   Molecular Formula: C12H17N2O4PS2
   Molecular Weight: 348.37
   No. of Ref: 2
   Data Present:
    Data  Ref
   +Ref/Only UDF    Data Type
       /2     PR Preparative Data
       2/     PB Physiological Data
Physiological Data
   Ecological Data
      Blattinsektizid (Ref. 2)
      Refs.
         2, Ruefenacht, et al, HCACAV, Helv.Chim.Acta, 59(1976)
            1593,1607

                              - end of display -
```

Searches in the /ED field can be used to explore possible relationships between, for example, Lawson Numbers and ecological properties:

```
?s ln=(29651 and 2826 and 2817) and dp=ed
            502   LN=29651
          62250   LN=2826
         196895   LN=2817
              6   DP=ED  (Ecological Data)
    S6        1   LN=(29651 AND 2826 AND 2817) AND DP=ED


?d /bi gs ed
        Display 6/BIGSED/1


881986
    Lawson No: 29651, 2826, 2817
    Beilstein Cit: 5-25
    Molecular Formula: C8H15N5
    Molecular Weight: 181.24
    Synonym: 4-Aethylamino-2-dimethylamino-5-amino-pyrimidin
    No. of Ref: 1
    'Graphic Structure:'
    881986
```

```
    Data Present:
     Data  Ref
     +Ref/Only UDF    Data Type
         /1    PR Preparative Data
         /1    CR Chemical Reactions
        1/     PB Physiological Data
   Physiological Data
     Ecological Data
        Rk. mit Mesoxalsaeurediaethylester, sd. n-NaHCO3-Lsg.--)
           6-Carbaethoxy-8-aethyl-2-dimethylamino-7(8H)-pteridinon
           (Ref. 1)
        Refs. 1, Pfleiderer, Taylor, JACSAT, J.Amer.Chem.Soc.,
               82(1960)3765,3772


                          - end of display -
```

3F3. Toxicity

Acute toxicity data are carried in the TX field, which is a freetext field, and is displayable or searchable in the usual manner:

```
?s dp=tx
        S7      543  DP=TX  (Toxicity)

?d /k/300-301
        Display 7/K/300

Physiological Data
   Toxicity
        Table I (Ref. 1)

                                - end of record -
?
        Display 7/K/301

Physiological Data
   Toxicity
        LD (50) (Ref. 1)

                                - end of record -
```

The **TX** field can be used to seek acute toxicity data for groups of related compounds:

```
?s barbitur?/cn and dp=tx
            2959  BARBITUR?/CN
             543  DP=TX  (Toxicity)
        S8     5  BARBITUR?/CN AND DP=TX

?d /bn cn gs tx
        Display 8/BNCNGSTX/1

306806
5-(2-acetoxy-propyl)-5-(1-methyl-butyl)-barbituric acid
   'Graphic Structure:'
   '306806'
```

```
Physiological Data
   Toxicity
      Table II (Ref. 1)
      Refs.
          1, Smissman, et al, JMCMAR, J.Med.Chem., 14(1971)853

                              - end of record -
```

3F4. Use

If a particular use or application of a chemical is mentioned in papers describing the chemical, that information is placed in the US field of the Beilstein Database. This is a freetext field which can be displayed or searched with words or phrases:

```
?s dp=us
      S9      583  DP=US  (Use)

?d /k/100
        Display 9/K/100

Physiological Data
   Use
       Aktivitaet gegenueber L1210, lymphoide Leukemie (Tab. VI,
       S. 89) (Ref. 1)

                              - end of record -
```

Compounds having specific uses can be retrieved with this field. In the example below, 41 compounds in use as dyestuffs (Farbstoff) are retrieved, and one of them is displayed:

```
?s farbstoff/us
      S10     41  FARBSTOFF/US

?d /cn sy gs us/20-21
        Display 10/CNSYGSUS/20

  Synonym: (8-Amino-naphthol-(1)-disulfonsaeure-(3.6))-
            2-azo-4-piazthiol
```

```
'Graphic Structure:'
634543
```

```
Physiological Data
   Use
      Eigg. als Farbstoff (Ref. 1)
      Refs.
         1, Algerino, et al, ANCRAI, Ann.Chim.(Rome), 50(1960)
            1703,1706,1707

                              - end of record -
?
      Display 10/CNSYGSUS/21

   Synonym:(1.8-Dihydroxy-naphthalin-disulfonsaeure-(3.6))-
           2-azo-5-piazthiol
   'Graphic Structure:'
   634542
```

```
Physiological Data
   Use
      Eigg. als Farbstoff (Ref. 2)
      Refs.
         2, Algerino, et al, ANCRAI, Ann.Chim.(Rome),
            50(1960)1703,1706,1707

                              - end of record -
```

3G. ELECTRICAL PROPERTIES

Most organic compounds do not conduct electricity freely and are therefore classified as dielectrics, or nonconductors. When a dielectric is introduced between the plates of

TABLE XIII

PROPERTY	SEARCH FIELD	DISPLAY FIELDS	
Dielectric constant	DIC	PP	EL
Dielectric static constant	DISC	PP	EL
Dielectric constant temperature	TEMP=	PP	EL
Dielectric static constant, temp.	TEMP=	PP	EL
Dielectric constant, frequency	DICF	PP	EL
Electrical Data	RTEL	PP	EL

a capacitor, the capacitance is increased by a factor *e*, called the dielectric constant. Dielectric constants can be easily measured experimentally and are related to the molecular polarizability (see section 3D6) and the dipole moment (see section 3D4) of the compound.

All the fields describing these electrical properties of molecules are summarized in Table XIII.

All these fields are numeric and can be searched in the usual manner.

3G1. Dielectric Constant

The *dielectric constant*, which, being a ratio, has no units, is carried in the DIC field of the Beilstein Database. The *temperature* (TEMP=) at which the dielectric constant is measured is important, as is the *electrical frequency* (DICF) that is used in the measurement. The dielectric constant is typically between 0 and 1000.

```
?s dic=0:50
        S1      568   DIC=0:50
```

Searches in the DIC field can be accompanied by temperature and frequency specifications:

```
?s dic=0:50(s)temp=20:30
           568    DIC=0 : DIC=50
        213814    TEMP=20 : TEMP=30
     S2    385    DIC=0:50(S)TEMP=20:30

?d /cn sy k/155-156
        Display 2/CNSYK/155

acetic acid-(2-bromo-1-methyl-propyl ester)
  Synonym: 3-Brom-2-acetoxy-butan
```

```
Electrical Properties
  Dielectric Constant: 7.41; Temp: 25 C; (Comment: (+-)- threo
  -3-bromo-2-acetoxy-butane) (Ref. 5 handbook...
  Dielectric Constant: 7.27; Temp: 25 C; (Comment: (+-)- erythro
  -3-bromo-2-acetoxy-butane) (Ref. 5 handbook)

                                    - end of record -


?
      Display 2/CNSYK/156


2-bromo-butyric acid
  Synonym: (+-)-2-Brom-buttersaeure
      racemische 2-Brom-buttersaeure
      inakt. alpha-Brom-buttersaeure
Electrical Properties
  Dielectric Constant: 7.22; Frequency: 500000000 Hz; Temp: 20 C;
  (Ref. 30 handbook)

                                    - end of display -


?s dic=0:50(s)dicf=1000:5000000000
            568   DIC=0 : DIC=50
            324   DICF=1000 : DICF=5000000000
      S3    195   DIC=0:50(S)DICF=1000:5000000000

?d /cn sy k/30-31
      Display 3/CNSYK/30


1,2,3-tribromo-propane
Electrical Properties
  Dielectric Constant: 6.45; Frequency: 1500000 Hz; Temp: 20 C;
  (Ref. 21 handbook...
  Dielectric Constant:  6.4; Frequency: 500000000 Hz; Temp: 19.8
  C; (Ref. 36 handbook)

                              - end of record -


?
      Display 3/CNSYK/31


1,2,3-trichloro-propane
Electrical Properties
  Dielectric Constant: 7.45; Frequency: 5.0E9 Hz; Temp: 21 C;
  (Ref. 32 handbook)

                              - end of record -
```

3G2. Dielectric Static Constant

The *dielectric static constant* —which is the dielectric constant measured with direct, as opposed to alternating, current—is stored in the DISC field, and the measurement temperature is in the TEMP= field. The dielectric static constant generally assumes very small values compared to the normal dielectric constant. The only measurement parameter of importance is the temperature, which can be used in searches:

```
?s disc=0:3(s)temp=10:20
            110   DISC=0 : DISC=3
         149491   TEMP=10 : TEMP=20
     S4      52   DISC=0:3(S)TEMP=10:20

?d /bi k/11
        Display 4/BIK/11

1736258
hexamethyl-disiloxane
  German Chem. Name: Hexamethyl-disiloxan
  Lawson No: 3793
  Beilstein Cit: 3-04-00-01859; 4-04-00-04018
  Molecular Formula: C6H18OSi2
  Molecular Weight: 162.38
  No. of Ref: 77
  Data Present:
   Data  Ref
   +Ref/Only UDF    Data Type
      10/       PR Preparative Data
      20/       CR Chemical Reactions
      75/17     PP Physical Properties
Electrical Properties
  Static Dielectric Constant: 2.17; Temp: 20 C; (Ref. 39
  handbook...

                        - end of record -
```

3H. MAGNETIC PROPERTIES

Many molecular magnetic properties are described in the Beilstein Database and are discussed in the section on spectroscopy (section 3M) of this manual. The only bulk magnetic property in the database is the magnetic susceptibility, and the fields of information associated with it are listed in Table XIV.

TABLE XIV

PROPERTY	SEARCH FIELD	DISPLAY FIELDS	
Magnetic susceptibility	MSUS	PP	MG
Magnetic susceptibility, temperature	TEMP=	PP	MG
Magnetic data	DP=RTMAG	PP	MG

3H1. Magnetic Susceptibility

When exposed to a magnetic field, most substances have a tendency to assume some magnetic character. The degree to which they do so depends upon the applied field and also upon the structure of the compound. The *magnetic susceptibility* of a compound is defined as the ratio of the magnetization produced to the intensity of the field to which it is subjected. The units of magnetic susceptibility are cm^3/mol, and the value is stored in the MSUS field. The magnetic susceptibilities of organic compounds are generally very small negative numbers, and accordingly, values for MSUS in the database are multiplied by 10^6.

For diamagnetic substances, the magnetic susceptibility is independent of temperature, but for paramagnetic compounds it is approximately proportional to the absolute temperature. For this reason, the *measurement temperature* is carried in the TEMP= field, and both of these numeric fields can be searched or displayed:

```
?s -120<msus<-119
     S1        5   -120<MSUS<-119

?d /k
     Display 1/K/1

Magnetic Properties
  Magnetic Susceptibility: -119.5 (cm**3/mol)*10E-6; (Ref. 1
  handbook)

                              - end of display -
```

It is useful to search for magnetic susceptibilities and for measurement temperature at the same time, and these terms should be linked by the proximity operator (S) to ensure that they are associated with one another in the database:

```
?s msus=-120:-110(s)temp=16:20
              36   MSUS=-120 : MSUS=-110
          143698   TEMP=16 : TEMP=20
     S2        2   MSUS=-120:-110(S)TEMP=16:20
```

Variation of magnetic susceptibility with structure can be studied by retrieving the magnetic susceptibilities for a series of related compounds:

```
?s (nitro or dinitro)/cn and coumarin/cn and dp=msus
        28233  NITRO/CN
         5217  DINITRO/CN
         3175  COUMARIN/CN
          467  DP=MSUS  (Magnetic Susceptibility)
    S3      4  (NITRO OR DINITRO)/CN AND COUMARIN/CN AND
               DP=MSUS

?t /gs k/1-4

 2/GSK/1
6-hydroxy-4-methyl-5,7-dinitro-coumarin
  Graphic Structure:
  '308119'
```

Magnetic Properties
  Magnetic Susceptibility: -111.9 (cm3/mol)*10E-6; (Ref. 3
   handbook)**

```
 2/GSK/2
7-hydroxy-4-methyl-8-nitro-coumarin
  Graphic Structure:
  '225171'
```

Magnetic Properties
  Magnetic Susceptibility: -106 (cm3/mol)*10E-6; (Ref. 9
   handbook)**

```
 2/GSK/3
7-hydroxy-4-methyl-6-nitro-coumarin
  Graphic Structure:
  '219724'
```

Magnetic Properties
  Magnetic Susceptibility: -105.5 (cm**3/mol)*10E-6; (Ref. 7
    handbook)

```
 2/GSK/4
6-hydroxy-4-methyl-5-nitro-coumarin
  Graphic Structure:
  '217367'
```

Magnetic Properties
  Magnetic Susceptibility: -105.6 (cm**3/mol)*10E-6; (Ref. 3
    handbook)

3H2. Magnetic Data

Other information concerning bulk magnetic properties is stored in the field RTMAG. In the example below, use of the field availability qualifier DP= (see section 3A2) leads to a reference to the magnetic moment of a manganese complex:

```
?s dp=rtmag
       S3     176   DP=RTMAG  (Magnetic Data)

?d /sy k/70

       Display 3/SYK/70
```

```
    Synonym: Tris(3,3',4'-trimethylpyrromethanato)mangan(III)
Magnetic Properties
    Magnetic moment (Ref. 1)

                              - end of display -
```

Specific terms can be sought in this field, as shown below:

```
?s "anisotropy of magnetic susceptibility"
      S4        57  "ANISOTROPY OF MAGNETIC SUSCEPTIBILITY"

?d /k/5

?
      Display 4/CNK/5

cyanogen chloride
Magnetic Properties
    Anisotropy of magnetic susceptibility (Comment: Beitrag des
    Van-Vleck-Terms zur magnetischen Susceptibilitaet.) (Ref. 95
    handbook)

                              - end of display -
```

3I. OPTICAL PROPERTIES

Throughout the history of organic chemistry, light has been used extensively in efforts to characterize compounds. Optical properties (as distinct from spectroscopic properties, which are described in section 3N), include refraction, rotation, rotatory dispersion, and circular dichroism. The fields involved are listed in Table XV.

3I1. Refractive Index

The *refractive index* of liquids, defined as the sine of the angle of incidence divided by the sine of the angle of refraction, is an easily measured and very reproducible constant which is useful for characterization purposes. Refractive indices, which are dimensionless and routinely measured to 5 decimal places, are stored in the RI field of

TABLE XV

PROPERTY	SEARCH FIELD	DISPLAY FIELDS		
Refractive index	RI =	PP	OP	RI
Refractive index, temperature	TEMP =	PP	OP	
Refractive index, solvent	SOLVENT =	PP	OP	
Refractive index, wavelength	WL =	PP	OP	
Optical rotatory power	ORP =	PP	OP	
Optical rotatory power, temperature	TEMP =	PP	OP	
Optical rotatory power, solvent	SOLVENT =	PP	OP	
Optical rotatory power, wavelength	WL =	PP	OP	
Optical rotatory power, type	—	PP	OP	
Mutarotation	MUT =	PP	OP	
Mutarotation, temperature	TEMP =	PP	OP	
Mutarotation, solvent	SOLVENT =	PP	OP	
Mutarotation, wavelength	WL =	PP	OP	
Mutarotation, type	—	PP	OP	
Optical rotatory dispersion	ORD =	PP	OP	
Optical rotatory dispersion, solvent	SOLVENT =	PP	OP	
Circular dichroism	CDIC =	PP	OP	
Circular dichroism, solvent	SOLVENT =	PP	OP	

the Beilstein Database, which can be searched with single numbers or with ranges:

```
?s ri=1.52870
     S1        32   RI=1.52870

?s ri=1.56:1.57
     S2      2790   RI=1.56:1.57
```

Experimental parameters upon which the value of the RI depends are the *measurement temperature* (TEMP=) and the *wavelength* (WL=) of the light used. Refractive indices are usually measured at room temperature (15–25 °C), and a favored wavelength is 589.00 nm, which is the wavelength of light produced by the common sodium lamp. If these parameters are used in searches for refractive indices, the number of retrievals drops markedly:

```
?s ri=1.5626:1.5627
     S3        37   RI=1.5626:1.5627

?s ri=1.5626:1.5627(s)wl=589.0
               37   RI=1.5626 : RI=1.5627
           216526   WL=589.0
     S4        33   RI=1.5626:1.5627(S)WL=589.0
```

The (S) operator must be used here to ensure that the two terms are in the same subfield. Use of AND in this case leads to three false drops:

```
?s ri=1.5626:1.5627 and wl=589.0
           37   RI=1.5626 : RI=1.5627
       216526   WL=589.0
    S5     36   RI=1.5626:1.5627 AND WL=589.0

?s ri=1.5626:1.5627(s)wl=589.0(s)temp=20
           37   RI=1.5626 : RI=1.5627
       216526   WL=589.0
       130884   TEMP=20
    S6     20   RI=1.5626:1.5627(S)WL=589.0(S)TEMP=20

?d/bi gs op/20

       Display 6/BIGSOP/20

109887
2,5-dichloro-thiophene
  German Chem. Name: 2,5-Dichlor-thiophen
  Lawson No: 16858
  Beilstein Cit: 4-17-00-00243; 0-17-00-00033; 2-17-00-00036;
  5-17
  Molecular Formula: C4H2Cl2S
  Molecular Weight: 153.03
  No. of Ref: 44
  Graphic Structure:
  '109887'
```

```
  Data Present:
   Data  Ref
   +Ref/Only UDF    Data Type
      5/5 .  PR Preparative Data
      3/1    CR Chemical Reactions
     48/20   PP Physical Properties
      4/     DR Characterization Derivative
       /6    KW Short File Keywords

Optical Properties
   Refractive Index: 1.5608; Wavelength: 589 nm; Temp: 19 C;
   (Ref. 17 handbook)
   Refractive Index: 1.5626; Wavelength: 589 nm; Temp: 20 C;
   (Ref. 16 handbook)
```

```
    Refractive Index: 1.5627; Wavelength: 589 nm; Temp: 20 C;
    (Ref. 18 handbook)
    Refractive Index: 1.5575; Wavelength: 589 nm; Temp: 30 C;
    (Ref. 10 handbook)
    Refractive Index: 1.5572; Wavelength: 589 nm; Temp: 20 C;
    (Ref. 16 handbook)
    Refractive Index: 1.5672; Wavelength: 527 nm; Temp: 20 C;
    (Ref. 16 handbook)
    Refractive Index: 1.588; Wavelength: 430.8 nm; Temp: 20 C;
    (Ref. 16 handbook)
    Refractive Index: 1.5578; Wavelength: 656.3 nm; Temp: 20 C;
    (Ref. 20)
    Refractive Index: 1.5766; Wavelength: 486.1 nm; Temp: 20 C;
    (Ref. 20)

                                -more-
?

    Display 6/BIGSOP/20


    Refractive Index: 1.5628; Wavelength: 589 nm; Temp: 20 C;
    (Ref. 20)
    Refractive Index: 1.5886; Wavelength: 430.8 nm; Temp: 20 C;
    (Ref. 20)
    Refractive Index: 1.5602; Wavelength: 589 nm; Temp: 25 C;
    (Ref. 11)
    Refractive Index: 1.5644; Wavelength: 589 nm; Temp: 20 C;
    (Ref. 5)
    Refs.
        5, Conde, et al, SYNTBF, Synthesis, (1976 )412
        10, Keswani, Freiser, JACSAT, J.Amer.Chem.Soc., 71 (1949)
            218
        11, Felloni, Pulidori, ANCRAI, Ann.Chim.(Rome),
            51(1961)1027,1030
        16, Coonradt, et al, JACSAT, J.Amer.Chem.Soc., 70 (1948)
            2564, 2567
        17, Steinkopf, Koehler, LACHDL, Liebigs Ann.Chem., 532
            (1937) 250, 264
        18, Campaigne, Le Suer, JACSAT, J.Amer.Chem.Soc., 70 (1948)
            415
        20, Jeffery, et al, JCSOA9, J.Chem.Soc., (1961)570

                                - end of display -
```

3l2. Optical Rotatory Power

Any compound that contains one or more chiral, or asymmetric, centers and that has been separated from its optical antipode has the potential to exhibit optical rotation. Many natural chemicals meet these two conditions. *Optical rotation* is the phenomenon in which the plane of polarization of polarized light is rotated as the beam passes through a solution of the compound. The degree of rotation is a characteristic of the compound and also depends upon the *solution concentration*, the *solvent*, the *temperature*, the *path length* of the cell, and the *wavelength* used.

When a compound is reported as possessing optical rotatory power, the value of the rotation is stored in the ORP field of the Beilstein Database. It usually ranges between 0 and 720°, although with complex macromolecules it can be much higher. The value, measured under standard conditions, has often been used in the characterization of compounds, particularly sugars, steroids, and amino acids.

Searching in the ORP field can be for single values or ranges:

```
?s orp=25:35
        S7     6950   ORP=25:35

?s orp=29:31
        S8     1694   ORP=29:31
```

Optical rotatory power is dependent upon a number of parameters. The wavelength (WL=) is usually that of the sodium D line (589.00 nm), and the length of the cell or polarimeter tube is usually 1 dm, but both may be varied, and such variations will affect the ORP. The choice of solvent (SOLVENT=) is important, as is the measurement temperature (TEMP=). Rotation data may be cited as the measured rotation α; as the specific rotation, which is $100\alpha/(\text{length} \times \text{concentration})$; or as the molar rotatory power, which is (specific rotation) $\times$ (molecular weight)/100. Which of these is used is indicated in the ORP record field (and also in the MUT record: see section 3I3), respectively, by alpha, (alpha), or <M>. Of these, (alpha) is by far the most commonly encountered. The concentration is carried in the record in units of g/100 ml.

```
?s orp=7(s)alpha
             290   ORP=7
          167021   ALPHA
      S9     285   ORP=7(S)ALPHA

?s orp=29:31(s)solvent=chcl3
            1694   ORP=29 : ORP=31
           46858   SOLVENT=CHCL3
     S10     464   ORP=29:31(S)SOLVENT=CHCL3
```

Using these fields in conjunction with a search for a specific ORP value greatly increases the convergence of the search:

```
?s orp=29:31(s)solvent=chcl3(s)temp=25
            1694   ORP=29 : ORP=31
           46858   SOLVENT=CHCL3
```

```
          48524  TEMP=25
   S11        48  ORP=29:31(S)SOLVENT=CHCL3(S)TEMP=25

?t /k/5

 11/K/5
Optical Properties
  Optical Rotatory Power: 29.7 deg; Wavelength: 589 nm; Type:
  (alpha); Solvent: CHCl3; Concentration: 1; Temp: 25 C; (Ref.
1)

?s orp=29:31(s)solvent=chcl3(s)temp=25(s)wl=589.0
        1694  ORP=29 : ORP=31
       46858  SOLVENT=CHCL3
       48524  TEMP=25
      216526  WL=589.0
   S12     48  ORP=29:31(S)SOLVENT=CHCL3(S)TEMP=25(S)WL=
                589.0
```

The wavelength 589.0 nm is used so commonly that it is not a particularly discriminatory search term. In this case it does not change the number of retrievals:

```
?t /bi gs k

 12/BIGSK/1
1730648
Hepta-O-acetyl-keto-D-glycero-L-manno-(2)octulose
  German Chem. Name:
Hepta-O-acetyl-keto-D-glycero-L-manno-(2)octulose
  Lawson No: 1155, 1142
  Beilstein Cit: 4-02-00-00384
  Molecular Formula: C22H30015
  Molecular Weight: 534.47
  No. of Ref: 1
  Graphic Structure:
  '1730648'
```

```
    Data Present:
     Data  Ref
     +Ref/Only UDF     Data Type
         1/      PR Preparative Data
         1/      PP Physical Properties
Optical Properties
   Optical Rotatory Power: 31 deg; Wavelength: 589 nm; Type:
   (alpha); Solvent: CHCl3; Concentration: c=4; Temp: 25 C; (Ref.
   1 handbook)
```

The concentration (in g/100 ml) and the type of rotational measurement reported can be specified in the search statements, but without qualifiers:

```
?s orp=7(s)wl=589(s)alpha(s)temp=21(s)solvent=chcl3(s)4
           290  ORP=7
        216526  WL=589
        167021  ALPHA
          8035  TEMP=21
         46858  SOLVENT=CHCL3
        550147  4
    S13        1  ORP=7(S)WL=589(S)ALPHA(S)TEMP=21
                  (S)SOLVENT=CHCL3(S)4


?t /bi gs op

 13/BIGSOP/1
1730325
penta-O-acetyl-keto-D-gluco-1-deoxy-(2)heptulose
   German Chem. Name:
Penta-O-acetyl-keto-D-gluco-1-desoxy-(2)heptulose
   Lawson No: 1155, 1124
   Beilstein Cit: 4-02-00-00372
   Molecular Formula: C17H24O11
   Molecular Weight: 404.37
   No. of Ref: 1
   Graphic Structure:
   '1730325'
```

```
   Data Present:
    Data  Ref
    +Ref/Only UDF    Data Type
        2/      PR Preparative Data
        2/      PP Physical Properties
Optical Properties
    Optical Rotatory Power: 7 deg; Wavelength: 589 nm; Type:
    (alpha); Solvent: CHCl3; Concentration: c=4; Temp: 21 C; (Ref.
    1 handbook)
    Refs.
       1, Wolfrom, et al, JACSAT, J.Amer.Chem.Soc., 79(1957)6454


?s orp=27.5:75.8(s)M

            22472  ORP=27.5 : ORP=75.8
            10780  M
      S14     479  ORP=27.5:75.8(S)M


?t /k/5,58-60


 14/K/5
Optical Properties
  Optical Rotatory Power: 35 deg; Wavelength: 589 nm; Type: (M);
  Solvent: aq. H2SO4; Temp: 20 C; (Comment: dextrorotatory form
  of the propane-dicarboxylic acid-(1...


 14/K/58
Optical Properties
  Optical Rotatory Power: 75 deg; Wavelength: 589 nm; Type: (M);
  Temp: 25 C; (Ref. 1)


 14/K/59
Optical Properties
  Optical Rotatory Power: 30.5 deg; Wavelength: 589 nm; Type:
  (M); Temp: 20 C; (Ref. 1...
  Optical Rotatory Power  34.5 deg; Wavelength: 589 nm; Type: (M
  ); Temp: 20 C; (Ref. 2)


 14/K/60
Optical Properties
  Optical Rotatory Power: 30.6 deg; Wavelength: 589 nm; Type: (M
  ); Temp: 20 C; (Ref. 1)
```

3l3. Mutarotation

Mutarotation is a process whereby the measured rotation of a compound in solution changes over time. This change results from the spontaneous, partial or complete conversion of one isomeric form to another, e.g. α-glucose to β-glucose. The final result is usually some mixture of the two forms, and the final rotation depends upon the composition of the mixture.

If a compound is reported to mutarotate, then the starting and final specific rotations will be entered into the MUT= field of the Beilstein Database, provided both are available. When only a single value is in this field, it is probably the final value. Usually, however, this entry will be a range, such as 27.5–75.8. It will be retrieved by any search statement which contains a number within that range or which is itself a range which overlaps with the database range. Consequently a simple search for a range in this field will inevitably produce many false drops because of range overlapping:

```
?s mut=27.5:75.8
      S15      402   MUT=27.5:75.8
```

False drops can be eliminated or minimized by successive addition of search criteria such as those citing the *solvent* (SOLVENT=), the *wavelength* (WL=), or the *temperature* (TEMP=):

```
?s mut=27.5:75.8(s)solvent=h2o
              402   MUT=27.5 : MUT=75.8
            55954   SOLVENT=H2O
      S16     292   MUT=27.5:75.8(S)SOLVENT=H2O

?s mut=27.5:75.8(s)solvent=h2o(s)wl=589.00
              402   MUT=27.5 : MUT=75.8
            55954   SOLVENT=H2O
           216526   WL=589.00
      S17     288   MUT=27.5:75.8(S)SOLVENT=H2O(S)WL=589.00

?s mut=27.5:75.8(s)solvent=h2o(s)wl=589(s)temp=20:21
              402   MUT=27.5 : MUT=75.8
            55954   SOLVENT=H2O
           216526   WL=589
           137521   TEMP=20 : TEMP=21
      S18     166   MUT=27.5:75.8(S)SOLVENT=H2O(S)WL=589
                    (S)TEMP=20:21

?t /k

 19/K/1
1729080
2-((Xi)-2-bromo-propionylamino)-2-deoxy-alpha-D-galactopyranose
   German Chem. Name:
2-((Xi)-2-Brom-propionylamino)-2-desoxy-alpha-D-galactopyranose
   Lawson No: 3354, 1165
   Beilstein Cit: 4-04-00-02048
```

Optical Properties

```
   Molecular Formula: C9H16BrNO6
   Molecular Weight: 314.13
   No. of Ref: 1
   Data Present:
    Data  Ref
    +Ref/Only UDF    Data Type
       1/      PR Preparative Data
       4/      PP Physical Properties
Optical Properties
  Mutarotation: 68.1 - 61.1 deg; Wavelength: 589 nm; Type:
  (alpha); Time: 480 - 1080 sec; Solvent: H2O; Concentration
  (c,p): c=1; Temp: 20 C; (Ref. 1 handbook...
  Mutarotation: 57.8 - 53.6 deg; Wavelength: 589 nm; Type:
  (alpha); Time: 1800 - 3600 sec; Solvent: H2O; Concentration
  (c,p): c=1; Temp: 20 C; (Ref. 1 handbook...
  Mutarotation: 50.7 - 49.5 deg; Wavelength: 589 nm; Type:
  (alpha); Time: 7200 - 14400 sec; Solvent: H2O; Concentration
  (c,p): c=1; Temp: 20 C; (Ref. 1 handbook)
```

The starting and ending rotations observed during the mutarotation can be used in search statements:

```
?s mut=(20:24 (s) 50:52)
             153   MUT=20 : MUT=24
              38   MUT=50 : MUT=52
      S19      13   MUT=(20:24 (S) 50:52)

?t/k/1-3

 19/K/1
Optical Properties
  Mutarotation  22 - 51 deg; Wavelength: 589 nm; Type: (alpha);
  Time: 120 sec; Solvent: H2O; Concentration (c,p...

 19/K/2
Optical Properties
  ...Mutarotation: -22.5 - 54.4 deg; Wavelength: 589 nm; Type:
  (alpha); Time: 600 sec; Solvent: H2O...

 19/K/3
Optical Properties
  ...Mutarotation: -15.8 - 61.2 deg; Wavelength: 589 nm; Type:
  (alpha); Time: 300 sec; Concentration (c...
```

Other mutarotation parameters include the time in seconds for mutarotation to be completed, the length of the tube used in the experiment (almost always 10 cm), and the concentration of the solution (in g/100 ml). These parameters are not searchable, but can be examined in individual records retrieved as described above.

314. Optical Rotatory Dispersion

The optical rotatory power of a compound historically was always measured at the *D* line of a sodium lamp (589.00 nm), and it is typically this value which is cited in connection with mutarotation, for example. It was discovered in 1895 by Cotton, however, that as the wavelength is varied, the rotation may change, and if the compound has an absorption band, the rotation will increase strongly, then fall off and change sign. The resulting curve of rotation versus wavelength, referred to as the *optical rotatory dispersion* (ORD) of the compound, will show the so-called *Cotton effect*. Cotton effects often shed light upon stereochemistry, and consequently much measurement in this area has been carried out between 1950 and the present.

The presence in a record of optical rotatory dispersion data is signified by the entry, in the ORD field, of the wavelength ranges of the data. Thus if an ORD curve is reported from 400 to 650 nm, then it will be retrieved by the search phrase ORD= 400:650. The actual data are not available online, and recourse must be had to the original paper to retrieve these.

```
?s ord=553:670
    S20      241   ORD=553:670

?t/k/1-3

 20/K/1
Optical Properties
  Optical Rotation Dispersion: 553 - 693 nm; Solvent: dioxane;
  (Ref. 2 handbook)

 20/K/2
Optical Properties
  Optical Rotation Dispersion: 254 - 670 nm; Solvent: CHCl3;
  (Ref. 10 handbook), (Ref. 12 handbook...
  Optical Rotation Dispersion: 423 - 656 nm; Solvent: CHCl3;
  (Ref. 10 handbook),  (Ref. 12 handbook

 20/K/3
Optical Properties
  Optical Rotation Dispersion: 284 - 643 nm; Solvent: ethyl
  acetate; (Ref. 4 handbook)
```

If a single number is used, only records containing that number (usually as the end of a reported range of wavelengths) will be retrieved. Ranges which bracket the number are not retrieved:

```
?s ord=553.00
      S21      1   ORD=553.00
```

but ranges which bracket the entered range are retrieved:

```
?s ord = 250
    S22     14  ORD = 250

?s ord = 250:251
    S23     16  ORD = 250:251

?c 23 not 22
              16  23
              14  22
      S24      2  23 NOT 22

?t/k/1-2

 24/K/1
Optical Properties
  Optical Rotation Dispersion: 251 - 670 nm; Solvent: CHCl3;
  (Ref. 1 handbook),  (Ref. 4 handbook...

 24/K/2
Optical Properties
  Optical Rotation Dispersion: 251 - 670 nm; Solvent: CHCl3;
  (Ref. 2 handbook...
  Optical Rotation Dispersion: 251 - 670 nm; Solvent:
  1,1,2,2-tetrachloro-ethane; (Ref. 2 handbook)
```

These two records were retrieved by ORD=250:251 but not by ORD=250.

Optical rotatory dispersion data depend mainly upon the solvent used. The identity of the solvent is in the SOLVENT= field. The wavelength is a variable, and other parameters such as concentration and cell length are standardized:

```
?s ord=443(s)solvent=chcl3
               2   ORD=443
           46858   SOLVENT=CHCL3
      S25      2   ORD=443(S)SOLVENT=CHCL3
```

```
?t/k/1-2

 25/K/1
Optical Properties
  Optical Rotation Dispersion: 443 - 640 nm; Solvent: CHCl3;
  (Ref. 8 handbook)

 25/K/2
Optical Properties
  Optical Rotation Dispersion: 443 - 644 nm; Solvent: CHCl3;
  (Ref. 2 handbook)
```

Use of a wavelength range in the search statement will give more retrievals:

```
?s ord=430:450(s)solvent=chcl3
            141   ORD=430 : ORD=450
          46858   SOLVENT=CHCL3
    S26      23   ORD=430:450(S)SOLVENT=CHCL3

?t /bi gs k/10-11

 26/BIGSK/10
98242
01,03;02,05;04,06-(R,R)-tribenzylidene)-D-mannitol
  German Chem. Name: 01,03;02,05;04,06-(R,R)-Tribenzyliden)-D-
  mannit
  Lawson No: 24049
  Beilstein Cit: 4-19-00-06244; 0-19-00-00464; 5-19
  Molecular Formula: C27H2606
  Molecular Weight: 446.5
  Synonym: 01,03;02,05;04,06-((R,R)-Tribenzyliden-D-mannit;
        (d-Mannit)-tribenzalaether
  No. of Ref: 17
  Graphic Structure:
  '98242'
```

```
   Data Present:
    Data  Ref
    +Ref/Only UDF    Data Type
       4/1     PR Preparative Data
      24/      PP Physical Properties
       /3      KW Short File Keywords
Optical Properties
    ...Optical Rotation Dispersion: 435.9 - 671.6 nm; Solvent:
       CHCl3; (Ref. 11 handbook)


 26/BIGSK/11
94259
(-)-Yohimbon
   German Chem. Name: (-)-Yohimbon
   Lawson No: 28737
   Beilstein Cit: 4-24-00-00654; 5-24
   Molecular Formula: C19H22N2O
   Molecular Weight: 294.4
   No. of Ref: 34
   Graphic Structure:
   '94259'
```

```
   Data Present:
    Data  Ref
    +Ref/Only UDF    Data Type
       8/8     PR Preparative Data
       4/7     CR Chemical Reactions
      19/      PP Physical Properties
       1/      DR Characterization Derivative
       /3      KW Short File Keywords
Optical Properties
    ...Optical Rotation Dispersion: 643.8 - 435.8 nm; Solvent:
       CHCl3; (Ref. 24 handbook)
```

315. Circular Dichroism

It transpires that left hand polarized light and right hand polarized light are absorbed unequally by optically active compounds having absorption bands in the 200–800 nm range. The result of this is that incident linearly polarized light, on passing through the sample, becomes elliptically polarized. This phenomenon is termed *circular dichroism*, and it has been used to characterize organic compounds. In circular dichroism studies, the variation with wavelength of the *molecular ellipticity* θ is measured. The entry in the CDIC field of the Beilstein Database is the wavelength range measured. The solvent used is identified in the SOLVENT= field, and both these fields may be used in searching:

```
?s cdic=200:300(s)solvent=methanol
            39   CDIC=200 : CDIC=300
         64300   SOLVENT=METHANOL
    S27     12   CDIC=200:300(S)SOLVENT=METHANOL

?t /bi k

 27/BIK/1
97272
ent-(16 E,19 R)-17,19-epoxy-15 beta H-coryn-16-ene-16-carboxylic
acid methyl ester; akuammigine
   German Chem. Name: ent-(16 E,19 R)-17,19-Epoxy-15 beta
   H-coryn-16-en-16-carbonsaeure-methylester; Akuammigin
   Lawson No: 32196, 289
   Beilstein Cit: 4-27-00-07928; 5-27
   Molecular Formula: C21H24N2O3
   Molecular Weight: 352.43
   No. of Ref: 21
   Data Present:
    Data  Ref
    +Ref/Only UDF    Data Type
       4/6      PR Preparative Data
       1/       CR Chemical Reactions
      20/2      PP Physical Properties
       /3       KW Short File Keywords
Optical Properties
   Circular Dichroism: 220 - 320 nm; Solvent: methanol; (Ref. 19
   handbook)
```

3J. ELECTROCHEMICAL PROPERTIES

A group of important physical properties are gathered together under the heading electrochemical. These properties, which are listed in Table XVI, deal mainly with the ionic dissociation of compounds and their oxidation–reduction potential (*redox* potential).

TABLE XVI

PROPERTY	SEARCH FIELD	DISPLAY FIELDS	
Dissociation exponent	DX =	EB	DX
Dissociation exponent, temperature	TEMP =	EB	DX
Dissociation exponent, solvent	SOLVENT =	EB	DX
Dissociation exponent, method	—	EB	DX
Dissociation exponent, type	—	EB	DX
Dissociation exponent, group	—	EB	DX
Enthalpy of dissociation	HDISC	EB	
Isoelectric point	IEP	EB	
Isoelectric point, solvent	SOLVENT =	EB	
Polarographic half wave potential	PHWPC	EB	
Redox potential	REDOXC	EB	

3J1. Dissociation Constant

Most organic acids are weak acids, and in solution they exist in equilibrium with their dissociated forms—generally a proton and the corresponding anion. The equilibrium constant (K_a) for this system is known as the *dissociation constant*. Similarly, weak organic bases exist in equilibrium with a protonated form and hydroxyl ion, and they too have a dissociation constant (K_b).

For weak electrolytes such as organic acids and bases, the un-ionized species dominates the equilibrium, with the result that the dissociation constants are very small numbers, typically between 10^{-3} and 10^{-14}. Because of this, it is customary to cite the negative base 10 logarithms of K_a or K_b: the pK_a and pK_b, respectively. This is the same as the logarithm of the reciprocal of the dissociation constant. It is this number that is called the *dissociation exponent* and stored in the DX field of the Beilstein Database:

```
?s dx=3.15
       S1      27   DX=3.15
```

In general, it is normal to search for a small range of pK values around the measured quantity:

```
?s dx=3.1:3.2
       S2     173   DX=3.1:3.2
```

and searches can be further converged if the various parameters of the measurement are cited. The measured value of a dissociation constant depends upon the temperature (TEMP=):

```
?s dx=3.1:3.2(s)temp=25
           173  DX=3.1 : DX=3.2
         48524  TEMP=25
    S3      81  DX=3.1:3.2(S)TEMP=25
```

The solvent used in dissociation studies most commonly is water, which is usually designated as h2o. In other solvents, compounds can give widely differing pK values, and it is thus wise to specify the solvent using the SOLVENT= field:

```
?s dx=3.1:3.2(s)temp=25(s)solvent=h2o
           173  DX=3.1 : DX=3.2
         48524  TEMP=25
         55954  SOLVENT=H2O
    S4      53  DX=3.1:3.2(S)TEMP=25(S)SOLVENT=H2O
```

Many methods exist for the measurement of dissociation constants, including such common ones as the potentiometric method, the conductometric method, and the spectrophotometric method. The method used for a particular measurement is carried in the dissociation constant record:

```
?s dx=3.1:3.2(s)temp=25(s)solvent=h2o(s)spectrophotometric
           173  DX=3.1 : DX=3.2
         48524  TEMP=25
         55954  SOLVENT=H2O
           955  SPECTROPHOTOMETRIC
    S5       4  DX=3.1:3.2(S)TEMP=25(S)SOLVENT=H2O(S)
                SPECTROPHOTOMETRIC

?t/bi gs k/1

 5/BIGSK/1
383619
nicotinic acid amide
  German Chem. Name: Nicotinsaeure-amid
  Lawson No: 26332
  Beilstein Cit: 4-22-00-00389; 0-22-00-00040; 2-22-00-00034;
  5-22
  Molecular Formula: C6H6N2O
  Molecular Weight: 122.13
  Synonym: Nicotinamid
  No. of Ref: 283
```

```
  Graphic Structure:
  '383619'
```

```
  Data Present:
   Data  Ref
   +Ref/Only UDF    Data Type
     24/39   PR Preparative Data
     71/22   CR Chemical Reactions
     78/91   PP Physical Properties
      1/     PB Physiological Data
      2/     DR Characterization Derivative
       /13   KW Short File Keywords
Electrochemical Behavior
   ...Dissociation Exponent: 3.1; Temp: 25 C; Solvent: H2O;
      Method: spectrophotometric; Type: a/apparent; (Ref. 125
      handbook...
```

Dissociation constants may be of several types. The two most common types are *thermodynamic* and *apparent* —words which are carried in the dissociation constant record. This record also carries a term (a1, a2, ..., b1, b2) which indicates that the pK in question is the first acidic pK, the second acidic, the first basic, and so on. Thus the term b1/thermodynamic means that it is the first basic pK, established thermodynamically:

```
?t 5/dx/1

 5/DX/1
Electrochemical Behavior
   Dissociation Exponent: 10.65; Temp: 20 C; Solvent: H2O;
   Method: spectrophotometric; Type: b1/thermodynamic; (Ref. 122
   handbook)
   Dissociation Exponent: 13.5; Temp: 20 C; Solvent: H2O; Method:
   spectrophotometric; Type: b2/thermodynamic; (Ref. 122
   handbook)
   Dissociation Exponent: 10.7; Temp: 25 C; Solvent: H2O; Method:
   potentiometric; Type: b1/apparent; (Ref. 123 handbook)
   Dissociation Exponent: 3.33; Temp: 20 C; Solvent: H2O; Method:
   spectrophotometric; Type: a/thermodynamic; (Ref. 124 handbook)
   Dissociation Exponent: 3.32; Temp: 20 C; Solvent: H2O; Method:
   potentiometric; Type: a/apparent; (Ref. 124 handbook)
   Dissociation Exponent: 3.12; Temp: 20 C; Solvent: H2O; Method:
   potentiometric; Type: a/apparent; (Ref. 124 handbook)
   Dissociation Exponent: 3.1; Temp: 25 C; Solvent: H2O; Method:
   spectrophotometric; Type: a/apparent; (Ref. 125 handbook)
   Dissociation Exponent: 3.15; Temp: 20 C; Solvent: aq. ethanol;
```

```
    Method: potentiometric; Type: a/apparent; (Ref. 126 handbook)
    Dissociation Exponent: (Ref. 127)
    Dissociation Exponent: (Ref. 128)
    Dissociation Exponent: (Ref. 129)
    Dissociation Exponent: (Ref. 130)
    Dissociation Exponent: (Ref. 131)
    Dissociation Exponent: (Ref. 132)
    Dissociation Exponent: (Ref. 133)
    Dissociation Exponent: (Ref. 134)
    Dissociation Exponent: (Ref. 135)
    Dissociation Exponent: (Ref. 136)
```

The term denoting type can be used in searches without qualifiers:

```
?s dx=3:4(s)a
          6896  DX=3 : DX=4
         31708  A
    S6     438  DX=3:4(S)A

?s dx=3:4(s)b1
          1142  DX=3 : DX=4
          3395  B1
    S7       3  DX=3:4(S)B1

?d/cn sy dx
      Display 7/CNSYDX/1

N, N'-di-tert-butyl-hexanediyldiamine
Electrochemical Behavior
    Dissociation Exponent: 3; Solvent: H2O; Method:
    potentiometric; Type: b1/apparent; (Ref. 1 handbook)
    Dissociation Exponent: 2.98; Solvent: H2O; Method:
    potentiometric; Type: b2/apparent; (Ref. 1 handbook)

                              - end of display -
```

The group that is dissociating is frequently identified in the record and can be searched for:

```
?s dx=8:9(s)"nh +"(s)temp=25(s)solvent=h2o(s)potentiometric
           902  DX=8 : DX=9
           146  NH +
         48524  TEMP=25
         55954  SOLVENT=H2O
          3968  POTENTIOMETRIC
    S8      10  DX=8:9(S)"NH +"(S)TEMP=25(S)SOLVENT=H2O(S)
                POTENTIOMETRIC
```

```
?t /bi gs dx/1

 8/BIGSDX/1
331267
4-(4-(4-chloro-phenyl)-4-hydroxy-piperidino)-1-(4-fluoro-phenyl)-
butan-1-one
   German Chem. Name: 4-(4-(4-Chlor-phenyl)-4-hydroxy-piperidino)-
   1-(4-fluor-phenyl)-butan-1-on
   Lawson No: 24823, 15505
   Beilstein Cit: 4-21-00-00657; 5-21
   Molecular Formula: C21H23ClFNO2
   Molecular Weight: 375.87
   No. of Ref: 33
   Graphic Structure:
   '331267'
```

```
   Data Present:
    Data   Ref
    +Ref/Only UDF    Data Type
       2/13    PR Preparative Data
        /1     CR Chemical Reactions
      21/17    PP Physical Properties
       6/      DR Characterization Derivative
        /2     KW Short File Keywords
Electrochemical Behavior
   Dissociation Exponent: 8.3; Dissociation Group: NH +; Temp: 25
   C; Solvent: H2O; Method: potentiometric; Type: a/apparent;
   (Ref. 2 handbook)
   Dissociation Exponent:  (Ref. 18)
   Dissociation Exponent:  (Ref. 19)
   Dissociation Exponent:  (Ref. 17)
   Dissociation Exponent:  (Ref. 23)
```

3J2. Enthalpy of Dissociation

The heat, or enthalpy, of dissociation of weak acids or bases is generally a very small quantity, typically $< 40,000$ J/mol. This is consistent with the insensitivity to temperature of dissociation constants. The enthalpy of dissociation is carried in the HDIS field of the Beilstein Database and temperature at which it was measured is in the record. Currently, the HDIS field must be searched with the dp=hdisc search statement:

```
?s dp=hdisc
     S9      118  DP=HDISC  (Enthalpy Of Dissociation
                            Comments/refs.)

?t/k/1-5

 9/K/1
.Electrochemical Behavior
   Enthalpy of Dissoc. (electrolytic): (Comment: at: 5-50 degree
   C.Enthalpie und Entropie der zweiten Dissoziationsstufe in...

 9/K/2
Electrochemical Behavior
   Enthalpy of Dissoc. (electrolytic): (Comment: in Wasser.)
   (Ref. 5 handbook)

 9/K/3
Electrochemical Behavior
   Enthalpy of Dissoc. (electrolytic): (Ref. 205 handbook)

 9/K/4
Electrochemical Behavior
   Enthalpy of Dissoc. (electrolytic): Temp: 25 C; (Comment: und
   Gibbs-Energie.) (Ref. 67 handbook)

 9/K/5
Electrochemical Behavior
   Enthalpy of Dissoc. (electrolytic): (Comment: at: 15-35
   degree C.und Entropie der Dissoziation.) (Ref. 11 handbook)
```

3J3. Isoelectric Point

Molecules such as amino acids or proteins have multiple charge centers and can act as acids or bases. Such molecules are electrolytic and will migrate in an electric field in a direction which is determined by their net charge. The net charge depends upon the pH of the medium. The *isoelectric point* for a compound is defined as that pH at which there is no net charge and therefore no migration of the molecule under the influence of the electric field. Isoelectric points are thus pH values and lie between 0 and 14. They are carried in the field IEP, and the solvent in which they are measured is identified in the SOLVENT= field:

```
?s iep=6:7
     S10        19   IEP=6:7

?d 10/cn k/1-2
         Display 10/CNK/1

N-dodecyl-glycine
Electrochemical Behavior
  Isoelectric Point (pH): 6.9; Solvent: H20; (Ref. 2 handbook)

                              - end of record -

?
      Display 10/CNK/2

N-glycyl-glycine
Electrochemical Behavior
  Isoelectric Point (pH): 6; (Ref. 121 handbook),  (Ref. 132
  handbook), (Ref. 133 handbook...

                              - end of display -
```

The solvent can be specified in the search statement:

```
?s iep=6:7(s)solvent=h2o
              19   IEP=6 : IEP=7
           55954   SOLVENT=H20
     S11      10   IEP=6:7(S)SOLVENT=H20

?d/cn sy k/4-5
         Display 11/CNSYK/4

DL-serine
Electrochemical Behavior
  Isoelectric Point (pH): 6; Solvent: H20; (Ref. 63 handbook)

                              - end of record -

?
      Display 11/CNSYK/5

DL-valine
Electrochemical Behavior
  Isoelectric Point (pH): 6; Solvent: H20; (Ref. 32 handbook)

                              - end of display -
```

3J4. Polarographic Half-Wave Potential

The current that passes between two electrodes in a cell containing a solution of a chemical depends upon the voltage difference between the two electrodes. When this potential difference is small, say less than 0.5 V, essentially no current flows. As the voltage is increased, there is at first no change in the current, but then the current increases sharply and plateaus at a level on the order of 10 μA. This *wave* is complete during a very small change of voltage, and the voltage at the midpoint of the wave is called the *half-wave potential*. The value of the half-wave potential depends upon the concentration of the solution, the nature of the solvent, and also the structure of the solute. It is therefore useful in the characterization of some organic compounds.

The polarographic half-wave potential, measured in volts, is stored in the PHWP field of the Beilstein Database. Currently, this field can be searched only with the dp=phwpc statement:

```
?s dp=phwpc
       S12    2480   DP=PHWPC   (Polarographic Half-wave Potential
                                 Com./refs.)

?d /k/200-201
        Display 12/K/200

Electrochemical Behavior
   Polarographic Half-wave Potential:  (Ref. 3)

                                - end of record -

?
        Display 17/K/201

Electrochemical Behavior
   Polarographic Half-wave Potential:  (Ref. 2)

                                - end of display -
```

3J5. Redox Potential

When a compound such as a quinone is mixed in solution with an equivalent amount of its reduced form (the hydroquinone), the solution will develop an electric potential which can be measured at an inert electrode. This potential, which is a measure of the proclivity of the system to form more quinone and less hydroquinone spontaneously, is called the *redox potential* of the system. Redox potentials are measured in volts and carried in the REDOX field of the Beilstein Database. The solvent in which they are measured is important and is identified in the record:

```
?s quinone/cn and dp=redoxc
            1127   QUINONE/CN
             575   DP=REDOXC  (Redox Potential Comments/refs.)
     S18      20   QUINONE/CN AND DP=REDOXC

?d /cn gs k/6
        Display 18/CNGSK/6

2,3-dimethyl-2,3-dihydro-naphtho(1,2-b)furan-4,5-quinone
2,3-dimethyl-2,3-dihydro-naphtho(1,2-b)furan-4,5- quinone
  Graphic Structure:
  '213125'
```

```
Electrochemical Behavior
    Redox Potential: (Comment: in waessrig-alkoholischer
    Salzsaeure in Gegenwart von Lithiumchlorid bei 25grad: 0,407
    V...

                              - end of display -
```

3K. Other Physical and Mechanical Properties

A variety of other physical and mechanical properties of chemical compounds are often measured and reported. These include *density* and *molar volume*, as well as *compressibility*, *coefficient of expansion*, *ultrasonic properties*, *surface tension*, and miscellaneous *mechanical properties*. All of these are stored in the Beilstein Database under the general heading of other physical and mechanical properties. The fields available in this section are listed in Table XVII.

TABLE XVII

PROPERTY	SEARCH FIELD	DISPLAY FIELDS		
Density	DN =	PP	PM	DN
Density, temperature	TEMP =	PP	PM	
Molar volume	MVOL =	PP	PM	
Molar volume, temperature	TEMP =	PP	PM	
Mechanical properties	DP = RTMEC	PP	PM	
Coefficient of expansion	DP = CEXPC	PP	PM	
Compressibility	DP = RTCOMP	PP	PM	
Ultrasonic properties	DP = RTUP	PP	PM	
Surface tension	ST =	PP	PM	ST
Surface tension, temperature	TEMP =	PP	PM	

3K1. Density

If the density of a compound is reported in the literature, it is carried in the DN field of the Beilstein Database. Densities, with units of g/cm^3, are expressed to 5 decimal places. The field may be searched with a single value:

```
?s dn=1.03280
        S1      21   DN=1.03280

?d /cn kwic
        Display 1/CNKWIC/1

1,5-diacetoxy-pentane
Other Physical & Mechanical Properties
   ...Density:  1.0328   g/cm**3; Measurement Temp: 15 C; (Ref. 12
handbook)

                             - end of display -
```

or a range:

```
?s dn=1.00:1.20
        S2    21144   DN=1.00:1.20
```

but truncation of the number entered is not allowed:

```
?s dn=1.03?
>>>Truncation not allowed on floating point data
```

   Density varies with temperature, and the temperature at which the density was
measured is carried in the TEMP= field. This can be searched in conjunction with the
density search using the (S) operator to ensure that the density and the temperature
belong to the same experiment:

```
?s dn=0.90:0.95(s)temp=20
            7704   DN=0.90 : DN=0.95
          130884   TEMP=20
     S3     4969   DN=0.90:0.95(S)TEMP=20
```

Use of the AND operator will lead to some false drops:

```
?s dn=0.90:0.95 and temp=20
            7704   DN=0.90 : DN=0.95
          130884   TEMP=20
     S4     5672   DN=0.90:0.95 AND TEMP=20
```

The 703 extra hits in S4 contain the correct density and temperature but not in the
same data line:

```
?c 4 not 3
            5672   4
            4969   3
     S5      703   4 NOT 3

?d /kwic/1-2
       Display 5/KWIC/1

Other Physical & Mechanical Properties
  Density:  0.9461  g/cm**3; Measurement Temp: 25 C; (Ref. 2
  handbook...
  Density: 0.9522 g/cm**3; Ref Temp: 4 C; Measurement Temp:
  20  C; (Ref. 3 handbook)
Optical Properties
  ...Refractive Index: 1.405; Wavelength: 589 nm; Temp:  20
C; (Ref. 3 handbook)

                         - end of record -
?
       Display 5/KWIC/2

Other Physical & Mechanical Properties
    ...Density: 0.9813 g/cm**3; Measurement Temp:  20  C; (Ref.
    2 handbook...
      Density:  0.9451  g/cm**3; Measurement Temp: 60 C; (Ref.
      2 handbook)
```

```
Viscosity
   ...Dynamic: 0.0771 g/(cm*sec); Temp:  20  C; (Ref. 2
      handbook...
Optical Properties
      Refractive Index: 1.4212; Wavelength: 589 nm; Temp:   20
  C; (Ref. 2 handbook)
                                - end of display -
```

3K2. Molar Volume

The *molar volume* of a compound is the volume occupied by one mole of the liquid. It is measured in cm³/mol and carried in the MVOL field. The measurement temperature is given in the TEMP= field, which should be linked to the MVOL value with the (S) operator so as to ensure that the two values are derived from the same experiment:

```
?s 45<mvol<49
        S6        1   45<MVOL<49

?d /cn kwic
         Display 6/CNKWIC/1

trifluoro-methane
Other Physical & Mechanical Properties
  Molar Volume: 48 cm**3; (Ref. 38 handbook)

                              - end of display -
?d /bi kwic
         Display 6/BIKWIC/1

1731035
trifluoro-methane
  German Chem. Name: Trifluor-methan
  Lawson No: 10
  Beilstein Cit: 3-01-00-00034; 4-01-00-00024; 0-01-00-00059
  Molecular Formula: CHF3
  Molecular Weight: 70.01
  No. of Ref: 82
  Data Present:
   Data  Ref
   +Ref/Only UDF    Data Type
     27/       PR Preparative Data
      9/       CR Chemical Reactions
     31/26     PP Physical Properties
Other Physical & Mechanical Properties
  Molar Volume:  48  cm**3; (Ref. 38 handbook)
                              - end of display -
```

```
?s mvol=80:90(s)temp=-253.1
                4    MVOL=80 : MVOL=90
               15    TEMP=-253.1
        S7      2    MVOL=80:90(S)TEMP=-253.1
?d /cn k/1-2
        Display 7/CNK/1

2-methyl-butane
Other Physical & Mechanical Properties
  Molar Volume: 85.1  cm**3; Temp:     -253.1 C; (Comment:
  Molvolumen::in unterkuehltem Zustand.) (Ref. 114
  handbook...Molar Volume: 82.7 cm**3; Temp:     -253.1 C;
  (Comment: Molvolumen::krystallisiert.) (Ref. 114 handbook...

                              - end of record -
?
        Display 7/CNK/2

pentane
Other Physical & Mechanical Properties
  Molar Volume: 81  cm**3; Temp:     -253.1 C; (Ref. 117
  handbook...

                              - end of display -
```

3K3. Mechanical Properties

A number of mechanical properties are stored in the RTMEC field, including virial
coefficients, *PVT* relationship data, and specific volume. This field can be searched
with the dp=rtmec search statement:

```
?s dp=rtmec
        S8      800   DP=RTMEC  (Mechanical Properties)

?t /k/50

 8/K/50
Other Physical & Mechanical Properties
    Mechanical Properties
        Specific volume (Ref. 6)
```

Specific terms can be introduced into the search statement, linked to this term with an (S) operator:

```
?s dp=rtmec(f)virial coefficients?
            800   DP=RTMEC  (Mechanical Properties)
            331   VIRIAL COEFFICIENTS?
      S9    150   DP=RTMEC(S)VIRIAL COEFFICIENTS?

?d /cn k/34
        Display 9/CNK/34

oxalonitrile
Other Physical & Mechanical Properties
   Mechanical Properties
       Virial coefficients of the equation of state (Ref. 113
       handbook)

                                    - end of record -
```

Note the ? indicating truncation of the longer term that is actually in the record.
    This field can be the basis of quite useful searches such as the one below, in which a reference to the virial coefficients of pent-1-ene was found:

```
?s pent-1-ene/cn and dp=rtmec(f)virial coefficients?
              1   PENT-1-ENE/CN
            800   DP=RTMEC  (Mechanical Properties)
            331   VIRIAL COEFFICIENTS?
            150   DP=RTMEC(S)VIRIAL COEFFICIENTS?
      S10     1   PENT-1-ENE/CN AND DP=RTMEC(F)VIRIAL COEFFICIENTS?
?d /k
        Display 10/K/1

 pent-1-ene
Other Physical & Mechanical Properties
   Mechanical Properties
       Virial coefficients of the equation of state (Ref. 173),
       (Ref. 174), (Ref...

                                    - end of display -
```

3K4. Coefficient of Expansion

Reports of measurements of the (thermal) coefficient of expansion of chemicals are cited in the CEXP field of the Beilstein Database. This is a comment field which can only be searched with the dp=cexpc search statement:

```
?s dp=cexpc
        S11    313  DP=CEXPC  (Coefficient Of Expansion
                                      Comments/refs.)

?d /cn sy k/1-2
        Display 11/CNSYK/1

  Synonym: p-Cyanobenzyliden-p-n-octyloxyanilin
Other Physical & Mechanical Properties
   Coefficient of Expansion: (Ref. 2)

                                   - end of record -
?
        Display 11/CNSYK/2

  Synonym: (p-n-Octyloxybenzylidene)-p-toluidine
Other Physical & Mechanical Properties
   Coefficient of Expansion: (Ref. 1)

                                   - end of display -
```

3K5. Compressibility

The compressibility, primarily of gases, is often reported, and with such compounds, this fact is noted in the RTCOMP field. This can be searched with the dp=rtcomp search statement:

```
?s dp=rtcomp
        S12    432  DP=RTCOMP  (Compressibility)

?t /k/60-61

 12/K/60
Other Physical & Mechanical Properties
   Compressibility
        Adiabatic compressibility (Ref. 16 handbook)

 12/K/61
Other Physical & Mechanical Properties
   Compressibility
        Isothermal compressibility (Ref. 96 handbook)
```

Specific terms can be linked to the dp= search statement with an (F) operator:

```
?s dp=rtcomp(F)adiabatic?
            432   DP=RTCOMP  (Compressibility)
            370   ADIABATIC?
      S13   326   DP=RTCOMP(F)ADIABATIC?

?d /cn k/300-301
        Display 13/CNK/300

furfuryl alcohol
Other Physical & Mechanical Properties
   Compressibility
       Adiabatic compressibility (Comment: aus der Schall-
       geschwindigkeit.) (Ref. 64 handbook)

                               - end of record -
?
        Display 13/CNK/301

2,6-dimethyl-pyridine
Other Physical & Mechanical Properties
   Compressibility
       Adiabatic compressibility (Ref. 120 handbook)

                               - end of display -
```

3K6. Ultrasonic Properties

Ultrasonic properties of chemicals, such as the velocity of sound in the material or its ability to support acoustic relaxation, are stored in the RTUP field. This field can be searched with the dp=rtup search statement:

```
?s dp=rtup
      S14   555   DP=RTUP  (Ultrasonic Properties)

?d /k/400-401
        Display 14/K/400

Other Physical & Mechanical Properties
   Ultrasonic Properties
       Velocity of sound (Ref. 407 handbook), (Ref. 408 handbook

                               - end of record -
```

```
?s dp=rtup(F)velocity of sound
            555   DP=RTUP  (Ultrasonic Properties)
            472   VELOCITY OF SOUND
     S15    468   DP=RTUP(F)VELOCITY OF SOUND


                              - end of display -
?d 15/bi k/3
        Display 15/BIK/3


2349369
  Lawson No: 250
  Beilstein Cit: 5-01
  Molecular Formula: C18H34
  Molecular Weight: 250.47
  No. of Ref: 2
  Data Present:
   Data  Ref
   +Ref/Only UDF    Data Type
       /1     PR Preparative Data
       /2     PP Physical Properties
Other Physical & Mechanical Properties
   Ultrasonic Properties
       Velocity of sound (Ref. 2)


                          - end of display -

?s dp=rtup(F)acoustic relaxation
            555   DP=RTUP  (Ultrasonic Properties)
            108   ACOUSTIC RELAXATION
     S16    /108  DP=RTUP(F)ACOUSTIC RELAXATION

?d 16/bi k/13
        Display 16/BIK/13


1730738
tetradeuterio-methane
  German Chem. Name: Tetradeuterio-methan
  Lawson No: 9
  Beilstein Cit: 3-01-00-00029; 4-01-00-00016; 5-01
  Molecular Formula: CD4
  Molecular Weight: 16.04
  No. of Ref: 404
```

```
  Data Present:
   Data  Ref
   +Ref/Only UDF    Data Type
      7/2    PR Preparative Data
    112/118  CR Chemical Reactions
     31/117  PP Physical Properties
      /16    KW Short File Keywords
 Other Physical & Mechanical Properties
    Ultrasonic Properties
         Acoustic relaxation (Ref. 76)

                        - end of display -
```

3K7. Surface Tension

A molecule in the body of a liquid is attracted equally in all directions by other molecules. If however the molecule is at the surface of the liquid, it experiences a net attraction towards the liquid, where there are more molecules than there are in the gas phase above the liquid. There is therefore a tendency for the molecules at the surface to contract, and this force is called *surface tension*.

The surface tension of a liquid is defined as the force (in dynes) acting at right angles to a line of unit length (1 cm) in the surface of the liquid. This datum is stored in the ST field of the Beilstein Database with the equivalent units of g/sec^2:

```
?s 50<st<100
       S17     19  50<ST<100

?d /kwic/9
       Display 17/KWIC/9

Other Physical & Mechanical Properties
       Surface Tension:  72.83  g/sec**2; Temp: 100 C; (Ref. 44
       handbook...
       Surface Tension:  68.64  g/sec**2; Temp: 130 C; (Ref. 44
       handbook)

                        - end of display -
```

The measurement temperature is carried in the TEMP= field, which can be searched with the ST field, the two search terms being linked with the (S) proximity operator:

```
?s st=50:100(s)temp=20
              19   ST=50 : ST=100
          130884   TEMP=20
     S18       6   ST=50:100(S)TEMP=20

?d /bi gs kwic/1-2
       Display 18/BIGSKWIC/1

1745235
bis-(2,3-dihydroxy-propyl)-ether
  German Chem. Name: Bis-(2,3-dihydroxy-propyl)-aether
  Lawson No: 636
  Beilstein Cit: 3-01-00-02327; 0-01-00-00513
  Molecular Formula: C6H14O5
  Molecular Weight: 166.17
  No. of Ref: 13
  Graphic Structure:
  '1745235'
```

```
  Data Present:
   Data  Ref
   +Ref/Only UDF    Data Type
      8/       PR Preparative Data
      3/       CR Chemical Reactions
     10/1      PP Physical Properties
Other Physical & Mechanical Properties
  Surface Tension:  53  g/sec**2; Temp:  20  C; (Ref. 10
handbook)

                          - end of record -
?
       Display 18/BIGSKWIC/2

1731048
tribromo-methane
  German Chem. Name: Tribrom-methan
  Lawson No: 10
  Beilstein Cit: 3-01-00-00088; 2-01-00-00033; 4-01-00-00082;
0-01-00-00068; 1-01-00-00016
  Molecular Formula: CHBr3
  Molecular Weight: 252.73
  No. of Ref: 314
  Graphic Structure:
  '1731048'
```

```
   Data Present:
    Data   Ref
    +Ref/Only UDF    Data Type
      22/      PR Preparative Data
      77/      CR Chemical Reactions
     244/66    PP Physical Properties
Other Physical & Mechanical Properties
   ...Surface Tension:  51  g/sec**2; Temp:  20  C; (Ref. 72
handbook...

                         - end of display -
```

TABLE XVIII

PROPERTY	SEARCH FIELD	DISPLAY FIELDS	
Heat capacity	CP =	PP	CA
Heat capacity, temperature	TEMP =	PP	CA
Heat capacity	CV =	PP	CA
Heat capacity, temperature	TEMP =	PP	CA
Heat capacity	CPO =	PP	CA
Heat capacity, temperature	TEMP =	PP	CA
Enthalpy of fusion	HMP =	PP	CA
Enthalpy of vaporization	HVP =	PP	CA
Enthalpy of vaporn., temp.	TEMP =	PP	CA
Enthalpy of sublimation	HSUB =	PP	CA
Enthalpy of sublimation, temp.	TEMP =	PP	CA
Enthalpy of phase transition	HPTP =	PP	CA
Enthalpy of formation	HFOR =	PP	CA
Enthalpy of formation, temp.	TEMP =	PP	CA
Enthalpy of combustion	HCOM =	PP	CA
Enthalpy of combustion, temp.	TEMP =	PP	CA
Entropy	DP = SDELTAC	PP	CA
Entropy of formation	SFOR =	PP	CA
Entropy of formation, temp.	TEMP =	PP	CA
Enthalpy of hydrogenation	HHYD =	PP	CA
Enthalpy of hydrogn., temp.	TEMP =	PP	CA
Hydrogenation product	—	PP	CA
Free energy of formation	GFOR =	PP	CA
Zero point energy	DP = EZPC	PP	CA

3L. THERMODYNAMIC PROPERTIES

Many thermodynamic, or caloric, properties of compounds are in the chemistry literature and have been collected in the Beilstein Database. As a matter of definition, these
are all concerned with energy, enthalpy, or entropy. They are listed in Table XVIII.

3L1. Heat Capacity

The amount of thermal energy necessary to raise the temperature of a material, per unit
temperature and per mole, is referred to as the *heat capacity* of the material. The heat
capacity of a substance can be measured either at constant pressure or at constant
volume. At constant pressure, the material is allowed to expand while being heated,
and the resulting heat capacity is termed C_P. If the material is not allowed to expand,
the heat capacity is C_V. Both quantities are cited in J/mol·K, and for gases they are
related to one another by the equation

$$MW \cdot (C_P - C_V) = R$$

where R is the ideal gas constant (8.314 J/mol·K) and MW is the molecular weight of
the compound. Thus the heat capacity of a gas at constant pressure is always greater
than that at constant volume. This is because work is required to expand the material.
Liquids and solids expand negligibly, and for them, C_P and C_V are approximately equal.

 Heat capacity varies with temperature, and accordingly, the measurement temperature is carried along with the values of C_P and C_V in the TEMP= field:

```
?s cp=100:140
        S1      206   CP=100:140

?s cp=100:140(s)temp=100:200
             206   CP=100 : CP=140
           47676   TEMP=100 : TEMP=200
        S2      29   CP=100:140(S)TEMP=100:200

?d /bi k/1
        Display 2/BIK/1

1736662
acetic acid methyl ester
  German Chem. Name: Essigsaeure-methylester
  Lawson No: 1155, 289
  Beilstein Cit: 3-02-00-00204; 4-02-00-00122; 2-02-00-00125;
  0-02-00-00124; 1-02-00-00052
  Molecular Formula: C3H6O2
  Molecular Weight: 74.08
  No. of Ref: 591
```

```
   Data Present:
    Data  Ref
    +Ref/Only UDF    Data Type
      71/      PR Preparative Data
      278/     CR Chemical Reactions
      380/93   PP Physical Properties
Heat Capacity
   cp: 115.14 J/(mol*deg); Temp: 136.9 C; (Ref. 92 handbook...

                            - end of display -
```

Searches for the heat capacity at constant volume, C_v, can be done in just the same way:

```
?s cv=180:190
        S3      1  CV=180:190

?d/bn cn k
        Display 3/BNCNK/1

124341
(1,3,5)triazine-2,4,6-triyltriamine
Heat Capacity
   cv: 185.87 J/(mol*deg); (Comment: cp ::bei 0 bis 80grad
   cal/Mol.) (Ref. 118 handbook)

                            - end of display -
```

The *heat capacity C_P^0* is cited along with the appropriate measurement temperature, which is in the TEMP= field:

```
?s cp0=385:390(s)temp=25
              2  CP0=385 : CP0=390
          48524   TEMP=25
      S4          CP0=385:390(S)TEMP=25

?d /k
      Display 4/K/1

Heat Capacity
   ...cp0: 143.19 - 385.6 J/(mol*deg); Temp: 25 - 1226.9 C;
   (Comment: cp:Hexan-Dampf.) (Ref. 164 handbook), (Ref. 165
   handbook...

                            - end of display -
```

3L2. Enthalpy

Enthalpy is a state function which for most purposes is equivalent to heat, which is not a state function. The *enthalpy* of a process is the amount of heat required for a unit amount of compound to undergo the process. Thus the *enthalpy of fusion*, also known as the latent heat of melting, is the thermal energy required to melt one gram of solid at the melting point. The molar enthalpy is the corresponding value for one mole of solid. In general, the Beilstein Database carries molar enthalpies rather than specific enthalpies.

The enthalpy of melting, in units of J/mol, is in the HMP field. Its values are all cited at the melting point, and no other relevant parameters exist. Searches in the HMP field can be carried out for single numbers or for ranges of numbers:

```
?s 9000<hmp<10000
     S5          28   9000<HMP<10000

?d /cn k
        Display 5/CNK/1

glutaronitrile
Calorific Data
  Enthalpy of Melting: 9545.9 J/mol; (Ref. 21 handbook)


                        - end of display -
```

The melting point is often cited together with the enthalpy of melting, and both can be obtained if CA (caloric data) is used in the display command. This, however, results in a lengthy listing of unwanted caloric data. To restrict the output to basic information (melting point and heat of melting), the set containing the compound (S6) should be intersected with a file retrieved with the dp=hmp search statement (S7), and then the resulting file (S8) can be displayed using cn, mp, and kwic in the format statement:

```
?s tetrahydro-furan/cn
     S6          1   TETRAHYDRO-FURAN/CN

?s dp=hmp
     S7        379   DP=HMP  (Enthalpy Of Melting)

?c 6 and 7
               1  6
             379  7
     S8          1  6 AND 7
```

```
?d /bi mp k
        Display 8/BIMPK/1


102391
Tetrahydro-furan
   German Chem. Name: Tetrahydro-furan
   Lawson No: 16789
   Beilstein Cit: 4-17-00-00024; 2-17-00-00015; 0-17-00-00010;
   1-17-00-00005; 5-17
   Molecular Formula: C4H8O
   Molecular Weight: 72.11
   No. of Ref: 1002
   Data Present:
    Data  Ref
    +Ref/Only UDF    Data Type
      54/27   PR Preparative Data
     193/188  CR Chemical Reactions
     209/179  PP Physical Properties
       1/     PB Physiological Data
        /22   KW Short File Keywords


                                     -more-
?
        Display 8/BIMPK/1


Crystals
   Melting Point:   -109 C; (Ref. 101 handbook)
   Melting Point:   -107.9 C; (Ref. 102 handbook)
   Melting Point:   -109.23 C; (Ref. 103 handbook)
   Melting Point:   -108.5 C; (Ref. 104 handbook)
   Melting Point:   -65 C; (Ref. 5 handbook)
   Melting Point:   -107.5 C; (Ref. 105)
   Melting Point:   -108.39 C; (Ref. 106)
   Refs.
       5, Bourguignon, CHZEA6, Chem.Zentralbl., 1908 I, 1630
     101, Brooks, Pilcher, JCSOA9, J.Chem.Soc., 1959 1535,1539
     102, Bissell, Finger, JOCEAH, J.Org.Chem., 24(1959)1259
     103, Critchfield, et al, JACSAT, J.Amer.Chem.Soc.,
          75(1953)6044
     104, Sisler, Mitarb, JACSAT, J.Amer.Chem.Soc., 70 (1948),
          3822
     105, Dyadin, et al, DKCHAY, Dokl.Chem.(Engl.Transl.),
          208(1973)9
     106, Lebedev, et al, JCTDAF, J.Chem.Thermodynamics,
          10(1978)321,323,324,326-328


                                     -more-
```

```
?
        Display 8/BIMPK/1

 Tetrahydro-furan
Calorific Data
    Enthalpy of Melting: 9106.3 J/mol; (Ref. 101 handbook)

                              - end of display -
```

The *enthalpy of vaporization*, HVP, is also in units of J/mol. This quantity is cited at specific temperatures and pressures, which are themselves carried in the TEMP= and PRES= fields, respectively:

```
?s hvp=40000:50000(s)temp=100
            223   HVP=40000 : HVP=50000
          11923   TEMP=100
    S9        1   HVP=40000:50000(S)TEMP=100

?d /bn cn gs k
        Display 9/BNCNGSK/1

105248
dihydro-furan-2-one
  Graphic Structure:
  '105248'
```

```
Calorific Data
    Enthalpy of Vaporization: 40884 J/mol; Temp: 100 C; Pressure:
    203 Torr; (Ref. 124 handbook)

                              - end of display -
```

The process of sublimation, in which a solid is converted to a gas without going through the liquid phase, is associated with an enthalpy. The *enthalpy of sublimation*, HSUB, is given in J/mol:

```
?s hsub=90000:100000
     S10       11   HSUB=90000:100000

?d /bn cn k
        Display 10/BNCNK/1
```

```
1759013
N-acetyl-glycine methylamide
Calorific Data
  Enthalpy of Sublimation: 97887 J/mol; (Comment: at:75-90
  degreeC.) (Ref. 11 handbook)

                              - end of display -
```

The temperature at which the enthalpy of sublimation was measured, when present, is in the TEMP= field. This may be searched, either alone or linked to the enthalpy by means of the (S) operator:

```
?s hsub=60000:80000(s)temp=0:100
            15   HSUB=60000 : HSUB=80000
        246830   TEMP=0 : TEMP=100
    S11       1   HSUB=60000:80000(S)TEMP=0:100

?d /bn cn k
        Display 11/BNCNK/1

136013
tetramethyl-(1,4)dioxane-2,5-dione
Calorific Data
  Enthalpy of Sublimation: 78712 J/mol; Temp: 25 C; (Ref. 4
  handbook)

                              - end of display -
```

3L3. Enthalpy of Phase Transition

Chemical compounds and mixtures often undergo phase transitions, distinct from melting and vaporization, under specific conditions of temperature, pressure, and (in the case of mixtures) composition. The *enthalpy of phase transition*, in J/mol, is stored in the HPT field of the Beilstein Database. This field can be searched either with single numbers or, more usefully, with a range of enthalpies:

```
?s hptp=3000:3100
    S12       1   HPTP=3000:3100

?t /bn cn k

 12/BNCNK/1
80580
benzo(b)thiophene
```

```
Calorific Data
   Other Phase Transition Enthalpies
        3012.8 J/mol; (Comment: bei 261.6 K.) (Ref. 81 handbook

?s dp=hptp
      S13      26  DP=HPTP  (Other Phase Transition Enthalpies)

?d /bi k
        Display 13/BIK/1

1795192
tricosanoic acid
  German Chem. Name: Tricosansaeure
  Lawson No: 1272
  Beilstein Cit: 3-02-00-01078; 4-02-00-01297; 2-02-00-00378
  Molecular Formula: C23H4602
  Molecular Weight: 354.62
  No. of Ref: 35
  Data Present:
   Data  Ref
   +Ref/Only UDF     Data Type
      8/        PR Preparative Data
     19/9       PP Physical Properties
      1/        DR Characterization Derivative
Calorific Data
   Other Phase Transition Enthalpies
        6782.6 J/mol; (Ref. 30 handbook)

                        - end of display -
```

3L4. Enthalpy of Formation

All chemicals possess an *enthalpy of formation*, which is the difference between the thermal energy of the molecule and that of its constituent atoms. This enthalpy, or heat of formation, measured in units of J/mol, is a basic and important property of molecules. It represents the heat that was absorbed by the system as the molecule was formed from its constituent atoms.

The enthalpy of formation depends upon the temperature (TEMP=) and pressure at which it is measured. Introduction into search statements of the temperature parameter, linked with the (S) operator, improves the precision of searches:

```
?s hfor=70000:1000000(s)temp=10:20
         73   HFOR=70000 : HFOR=1000000
     149491   TEMP=10 : TEMP=20
   S14     2  HFOR=70000:1000000(S)TEMP=10:20
```

```
?d /cn k
        Display 14/CNK/1

2-iodo-propane
Entropy & Enthalpy
  Enthalpy of Formation: 77037 J/mol; Temp: 20 C; (Comment:
  fluess..) (Ref. 46 handbook)

                        - end of display -
```

3L5. Enthalpy of Combustion

The *enthalpy of combustion* of a compound is the thermal energy released when a unit quantity of the compound in its standard state at 298.15 K is completely combusted in the presence of excess oxygen. This enthalpy is fairly easy to measure experimentally and frequently serves as the basis for calculations which lead to the enthalpy of formation of compounds. International convention has it that heat released by a system is negative, and therefore enthalpy of combustion is a negative number. It is measured in J/mol and is stored in the HCOM field of the Beilstein Database:

```
?s hcom=-4800000:-4700000
        S15     7  HCOM=-4800000:-4700000
```

The enthalpy of combustion depends upon the measurement temperature (TEMP=) and pressure (PRES=), and when these parameters are cited, the searches become more precise:

```
?s hcom=-1270000:-1260000(s)temp=20.0
            1   HCOM=-1270000 : HCOM=-1260000
       130884  TEMP=20.0
     S16     1   HCOM=-1270000:-1260000(S)TEMP=20.0

?d /bn cn gs k
        Display 16/BNCNGSK/1

102378
oxirane
  Graphic Structure:
  '102378'

                        △0

Entropy & Enthalpy
  Enthalpy of Combustion: - 1265700 J/mol; Temp: 20 C; (Ref. 69
  handbook...

                        - end of display -
```

```
?s hcom=-6300000:-6200000(s)temp=20(s)pres=700:760
              3   HCOM=-6300000 : HCOM=-6200000
         130884   TEMP=20
            684   PRES=700 : PRES=760
     S17      1   HCOM=-6300000:-6200000(S)TEMP=20(S)
                  PRES=700:760
?d /k
       Display 17/K/1


Entropy & Enthalpy
  Enthalpy of Combustion: - 6271800 J/mol; Temp: 20 C;
  Pressure: 735.51 Torr; (Comment: C8H2205Si2 fluessig -) CO2
  gasfoermig + H2O fluessig + SiO2 fest .) (Ref. 2 handbook)

                          - end of display -
```

3L6. Entropy

Entropy is a measure of the disorder in a system and is a fundamental thermodynamic
property whose relationship to enthalpy, temperature, and Gibbs free energy is determined by the second law of thermodynamics. Entropy is measured indirectly, cited in
$J/mol \cdot K$, and stored in the SDELTA field. Currently, this field must be searched with the
dp=sdeltac search statement:

```
?s dp=sdeltac
     S18     561   DP=SDELTAC  (Entropy Delta S Comments/refs.)

?t /k/1-5

 18/K/1

Entropy & Enthalpy
   Entropy Delta S: (Comment: Umwandlungsentropie.) (Ref. 54
   handbook)

 18/CNK/2
1-bromo-triacontane
Entropy & Enthalpy
   Entropy Delta S: (Comment: Umwandlungsentropie.) (Ref. 2
   handbook), (Ref. 3 handbook

 18/CNK/3
tritriacontane
Entropy & Enthalpy
   Entropy Delta S: (Ref. 12 handbook)
```

```
 18/CNK/4
4,5-dithia-octanedioic acid
Entropy & Enthalpy
   Entropy Delta S: (Comment: und freie Energie bei 298.1
   gradK.) (Ref. 15 handbook),  (Ref...

 18/CNK/5
octacosane
Entropy & Enthalpy
   Entropy Delta S: (Comment: Umwandlungsentropie.) (Ref. 46
   handbook), (Ref. 47 handbook
```

3L7. Entropy of Formation

The *entropy of formation* of a compound is equal to its enthalpy of formation, in J/mol, divided by the temperature. This quantity is stored in the SFOR field of the Beilstein Database, and the measurement temperature is stored in the TEMP= field:

```
?s sfor=181
       S19       1   SFOR=181

?d /k
       Display 19/K/1

Entropy & Enthalpy
  Entropy of Formation: 181 J/(mol*deg); (Comment: Standard.)
  (Ref. 87 handbook)

                              - end of display -
```

Search statements can include ranges of entropies and temperatures:

```
?s sfor=250:300(s)temp=0
              3   SFOR=250 : SFOR=300
           8897  TEMP=0
    S20       1   SFOR=250:300(S)TEMP=0

?d /bi gs k
       Display 20/BIGSK/1

103233
pyridine
  German Chem. Name: Pyridin
  Lawson No: 24225
```

```
Beilstein Cit: 4-20-00-02205; 2-20-00-00096; 0-20-00-00181;
1-20-00-00054; 5-20
Molecular Formula: C5H5N
Molecular Weight: 79.1
No. of Ref: 4724
Graphic Structure:
'103233'
```

```
Data Present:
 Data  Ref
 +Ref/Only UDF    Data Type
   92/46   PR Preparative Data
 1062/  791CR Chemical Reactions
 1049/  885PP Physical Properties
    4/      PB Physiological Data
   46/      DR Characterization Derivative


                              -more-
?

      Display 19/BIK/1


      /61   KW Short File Keywords
Entropy & Enthalpy
  ...Entropy of Formation: 276 J/(mol*deg); Temp: 0 C; (Ref.
     131 handbook...


                        - end of display -


?s sfor=200:500(s)temp=25:30
               6  SFOR=200 : SFOR=500
          64818  TEMP=25 : TEMP=30
       S21      2  SFOR=200:500(S)TEMP=25:30


?t /bi k/2

 21/BIK/2
1159
pyrrole
  German Chem. Name: Pyrrol
  Lawson No: 24173
  Beilstein Cit: 4-20-00-02072; 0-20-00-00159; 1-20-00-00036;
  2-20-00-00079; 5-20
  Molecular Formula: C4H5N
  Molecular Weight: 67.09
  No. of Ref: 930
```

```
   Data Present:
    Data  Ref
    +Ref/Only UDF    Data Type
      43/12    PR Preparative Data
     196/127   CR Chemical Reactions
     234/245   PP Physical Properties
        /17    KW Short File Keywords
 Entropy & Enthalpy
   Entropy of Formation: 272 J/(mol*deg); Temp: 26.9 C; (Ref. 142
   handbook)
```

3L8. Hydrogenation

The reduction of organic molecules with hydrogen gas in the presence of a catalyst has traditionally been a very important reaction in organic chemistry, and data concerning hydrogenation receive special treatment in the Beilstein Database. The enthalpy change associated with the complete hydrogenation of a compound is termed the *enthalpy of hydrogenation*, has units of J/mol and is stored in the HHYD field of the Beilstein Database. The temperature at which hydrogenation takes place is recorded in the record and can be retrieved with the TEMP= qualifier:

```
?s hhyd=150000:155000
      S22      1  HHYD=150000:155000

?d /cn k
        Display 22/CNK/1

Furan
Entropy & Enthalpy
  Enthalpy of Hydrogenation: 153360 J/mol; Temp: 81.9 C; (Ref.
  148 handbook)

                          - end of display -
```

The hydrogenation temperature can be used in the search statement with the TEMP= qualifier:

```
?s hhyd=-119000:-118000(s)temp=20:30
            1  HHYD=-119000 : HHYD=-118000
        213814  TEMP=20 : TEMP=30
    S23     1  HHYD=-119000:-118000(S)TEMP=20:30
```

The chemical name of the hydrogenation product is carried in the record and may be searched for:

```
?s dp=hhyd(s)isobutyric acid
              10   DP=HHYD   (Enthalpy Of Hydrogenation)
             278   ISOBUTYRIC ACID
      S24      1   DP=HHYD(S)ISOBUTYRIC ACID

?d /k
      Display 24/CNK/1

Entropy & Enthalpy
   Enthalpy of Hydrogenation: -118240 J/mol; Product name:
   isobutyric acid; Temp: 25 C; (Ref. 58 handbook)

                           - end of display -
```

3L9. Gibbs Free Energy

Any chemical compound has a given energy which, in principle, can be utilized for the accomplishment of work. This energy is termed free energy, to distinguish it from internal (nuclear or electronic) energy, and the name that is now generally used for it is *Gibbs free energy*, after the physical chemist who formulated much of this theory.

The Gibbs free energy *G* is related to the enthalpy *H*, the temperature *T*, and the entropy *S* by the well-known equation which restates the second law of thermodynamics:

$$G = H - TS.$$

This equation is true for any system, such as a liter of a gas or a mixture of liquids. More significantly, it is true for any process, such as a phase change (melting of a solid) or the formation of a chemical from its constituent elements.

The arbitrary assumption is made that the Gibbs free energy of all elements in their normal states is zero. The free energy of formation of a compound therefore is the energy absorbed or released as the compound is formed from elements. A compound that is rich in free energy may generally be expected to be reactive.

When a compound is formed from its elements, the free energy of formation (dG) is related to the enthalpy of formation (dH) and the entropy of formation (dS) by the equation

$$dG = dH - TdS,$$

where *T* is the (fixed) temperature under consideration.

The *Gibbs free energy of formation* for compounds is carried in the GFOR field of the Beilstein Database. It is given in units of J/mol, and the temperature and pressure of the measurement are also cited:

```
?s gfor=-334200:-334100
      S25        1  GFOR=-334200:-334100

?d 25/bi k
        Display 25/BIK/1

118515
phthalic acid-anhydride
  German Chem. Name: Phthalsaeure-anhydrid
  Lawson No: 18408
  Beilstein Cit: 0-17-00-00469; 4-17-00-06135; 2-17-00-00463;
  1-17-00-00251; 5-17
  Molecular Formula: C8H4O3
  Molecular Weight: 148.12
  No. of Ref: 915
  Data Present:
   Data  Ref
   +Ref/Only UDF    Data Type
     47/46   PR Preparative Data
    436/89   CR Chemical Reactions
    112/70   PP Physical Properties
       /13   KW Short File Keywords
Entropy & Enthalpy

                                    -more-

      Display 1/BIK/1

  Gibbs-Energy of Formation: -334110 J/mol; (Comment: Standard.)
  (Ref. 182, handbook)
                        -end of display-
```

3L10. Zero Point Energy

All molecules vibrate, and the energy associated with vibration is a component of the total energy of the molecule. The *zero point energy*, or residual energy, of a molecule is the energy which it must have at absolute zero (0 K) when all vibration ceases. Zero point energies are carried in the EZP field of the Beilstein Database and are cited in J/mol. Currently, this field must be searched with the dp=ezpc search statement:

```
?s dp=ezpc
      S26       11  DP=EZPC  (Zero-point Energy Comments/refs.)

?d /cn k/1-2
        Display 26/CNK/1
```

```
but-2 t-ene
Entropy & Enthalpy
   Zero-point Energy: (Comment: spektroskopisch ermittelt.)
   (Ref. 65 handbook), (Ref. 66 handbook

                        - end of record -
```

3M. TRANSPORT PROPERTIES

A number of properties of materials are essentially transport phenomena. These
include properties, such as viscosity and diffusion, that are generally temperature
dependent. The complete list of the fields dealing with transport properties is provided
in Table XIX.

TABLE XIX

PROPERTY	SEARCH FIELD	DISPLAY FIELDS	
Dynamic viscosity	VISD =	PP	TP
Dynamic viscosity, temperature	TEMP =	PP	TP
Kinematic viscosity	VISK =	PP	TP
Kinematic viscosity, temperature	TEMP =	PP	TP
Bulk viscosity	VISB =	PP	TP
Bulk viscosity, temperature	TEMP =	PP	TP
Thermal conductivity	DP = TCONC	PP	TP
Self-diffusion	DP = SDIFC	PP	TP

3M1. Dynamic Viscosity

The *dynamic viscosity* of a material is the force per unit area required to maintain a unit
difference of velocity between two layers a unit distance apart. Dynamic viscosity data
are stored in the VISD field of the Beilstein Database and measured in units of
$dyn \cdot sec/cm^2$ (poise). This is equivalent to $g/cm \cdot sec$, which are the units cited.
Dynamic viscosity depends upon temperature, and the measurement temperature can
be searched for with the TEMP= qualifier:

```
?s visd=.1:.2(s)temp=20:30
           275   VISD=.1 : VISD=.2
        213814   TEMP=20 : TEMP=30
    S1     144   VISD=.1:.2(S)TEMP=20:30
```

```
?d/cn sy k/20-22
        Display 1/CNSYK/20

pentanediyl-bis-phosphonic acid tetraethyl ester
Viscosity
  Dynamic: 0.1549 - 0.0433  g/(cm*sec); Temp: 30 - 70  C; (Ref.
  1 handbook)

                                - end of record -
?
        Display 1/CNSYK/21

adipic acid bis-(2-butoxy-ethyl ester)
Viscosity
  Dynamic: 0.125 g/(cm*sec); Temp: 20 C; (Ref. 2 handbook)


                                - end of record -
?
        Display 1/CNSYK/22

2-lauroyloxy-propionic acid-(2-ethyl-hexyl ester)
Viscosity
  Dynamic: 0.1627 g/(cm*sec); Temp: 20 C; (Ref. 1 handbook)


                                - end of record -
```

3M2. Kinematic Viscosity

The *kinematic viscosity* of a material is the ratio of its absolute (bulk) viscosity to its relative density. Kinematic viscosities are carried in the VISK field of the Beilstein Database. They are measured in units of $(g/cm \cdot sec)/(g/cm^3)$, which simplifies to cm^2/sec, the cited units. The measurement temperature is carried in the TEMP= field:

```
?s visk=.2:.3(s)temp=30:40
            58   VISK=.2 : VISK=.3
         12920   TEMP=30 : TEMP=40
     S2     18   VISK=.2:.3(S)TEMP=30:40

?d /bi gs k/3
        Display 2/BIGSK/3
```

```
1810572
hexadecaethyl-heptasiloxane
  German Chem. Name: Hexadecaaethyl-heptasiloxan
  Lawson No: 3794
  Beilstein Cit: 4-04-00-04163
  Molecular Formula: C32H8006Si7
  Molecular Weight: 757.58
  No. of Ref: 4
  Graphic Structure:
  '1810572'
```

```
  Data Present:
   Data  Ref
   +Ref/Only UDF    Data Type
      9/      PP Physical Properties
Viscosity
   ...Kinematic: 0.239 cm**2/sec; Temp: 38 C; (Ref. 1 handbook),
     (Ref. 4 handbook

                        - end of display -
```

3M3. Bulk Viscosity

The bulk viscosity of materials, in units of g/cm·sec, is stored in the VISB field. This field can be searched with single numbers or with ranges:

```
?s visb=0:.1
     S3      19  VISB=0:.1


?s visb=0:.1(s)temp=20:30
           19  VISB=0 : VISB=.1
       213814  TEMP=20 : TEMP=30
     S4      14  VISB=0:.1(S)TEMP=20:30
```

```
?d /cn tp/1-2
        Display 4/CNTP/1

heptadecanoic acid ethyl ester
Viscosity
    Bulk: 0.05815 g/cm/sec; Temp: 30 C; (Ref. 10 handbook)
    Bulk: 0.05095 g/cm/sec; Temp: 35 C; (Ref. 10 handbook)
    Refs.
        10, Mumford, Phillips, JCSOA9, J.Chem.Soc., 1950 75,79

                                - end of record -
?
        Display 4/CNTP/2

2,4-dimethyl-tetradecanoic acid
Viscosity
    Bulk: 0.6144 - 0.0477 g/cm/sec; Temp: 20 - 100 C; (Ref. 2
    handbook)
    Refs.
        2, Petrow, et al, FTSEAK, Fette Seifen, 61(1959)940,941,942

                                - end of record -
```

3M4. Thermal Conductivity

A measure of the ease with which heat is conducted from one point to another through a material is known as the *thermal conductivity* of the material. It is measured in $J/cm \cdot sec \cdot K$, and the data are carried in the TCON field of the Beilstein Database, which must be searched with the dp=tconc search statement:

```
?s dp=tconc
        S5      257  DP=TCONC  (Thermal Conductivity Comments/refs.)

?d /cn tp/25-26
        Display 5/CNTP/25

tetrafluoro-ethene
Self-Diffusion & Thermal Conductivity
    Thermal Conductivity: (Ref. 23 handbook)
    Refs.
        23, Renfrew, Lewis, IECHAD, Ind.Eng.Chem., 38(1946)870

                                - end of record -
```

```
  ?

          Display 5/CNTP/26

  octadecane
  Viscosity
     Dynamic: 0.03209 g/(cm*sec); Temp: 37.8 C; (Ref. 38 handbook)
     Dynamic: 0.02062 g/(cm*sec); Temp: 60 C; (Ref. 38 handbook)
     Dynamic: 0.0115 g/(cm*sec); Temp: 98.9 C; (Ref. 38 handbook)
     Dynamic: 0.0389 - 0.002 g/(cm*sec); Temp: 30 - 315 C; (Ref. 61
     handbook)
     Dynamic: 0.0356 g/(cm*sec); Temp: 32 C; (Ref. 60 handbook)
     Dynamic: 0.0279 g/(cm*sec); Temp: 42 C; (Ref. 60 handbook)
     Refs.
        38, Schiessler, et al, Am.Doc.Inst.Doc.4597
        60, Dover, Hensley, IECHAD, Ind.Eng.Chem., 27(1935)338
        61, Rossini, Selected Values 1953, S.228
     Kinematic: 0.009508 cm**2/sec; Temp: 149 C; (Ref. 38 handbook)
     Kinematic: 0.006279 cm**2/sec; Temp: 204 C; (Ref. 38 handbook)
     Kinematic: 0.005263 cm**2/sec; Temp: 232 C; (Ref. 38 handbook)
     Kinematic:   Temp: 37.8 - 99 C; (Ref. 43 handbook)
     Refs.
        38, Schiessler, et al, Am.Doc.Inst.Doc.4597


                                   -more-
  ?

          Display 5/CNTP/26


        43, Schiessler, Mitarb, Pr.am.Petr.Inst., 26 III(1946)269,
           CA (1948)1181
  Self-Diffusion & Thermal Conductivity
     Self-diffusion:  Temp: 40 - 170 C; (Ref. 62 handbook)
     Refs.
        62, Douglass, McCall, JPCHAX, J.Phys.Chem., 62(1958)1102,
           1106
     Thermal Conductivity: Temp: 15 - 55 C; (Ref. 63 handbook)
     Thermal Conductivity: Temp: 33 - 74 C; (Ref. 64 handbook)
     Refs.
        63, Sutherland, et al, IECHAD, Ind.Eng.Chem., 51(1959)585,
           587
        64, Sakiadis, Coates, A.I.Ch.E.Journal, 3(1957)121,123

                                 - end of record -
```

3M5. Self-Diffusion

Data concerning *self-diffusion*, i.e. movement of molecules within pure compounds or mixtures, is stored in the SDIF field. The self-diffusion coefficient has the units

cm^2/sec. Currently, this field may only be searched with the dp=sdifc search statement:

```
?s dp=sdifc
     S6        60  DP=SDIFC  (Self-diffusion Comments/refs.)

?d /cn tp/30-32
       Display 6/CNTP/30


propan-1-ol

Viscosity
   Dynamic: 0.563 - 0.03825 g/(cm*sec); Temp: -70.6 - -0.3 C;
   (Ref. 114 handbook)
   Dynamic: 0.386 - 0.0226 g/(cm*sec); Temp:  -60 - 20 C; (Ref.
   115 handbook)
   Dynamic: 0.02522 g/(cm*sec); Temp: 15 C; (Ref. 103 handbook)
   Dynamic: 0.01722 g/(cm*sec); Temp: 30 C; (Ref. 103 handbook)
   Dynamic: 0.01967 g/(cm*sec); Temp: 25 C; (Ref. 116 handbook)
   Dynamic:  (Comment: dynamic viscosity:bei Raumtemp. und
   Drucken bis zu 30000 atm.) (Ref. 117 handbook)
   Dynamic:  (Ref. 118 handbook),  (Ref. 119 handbook)
   Dynamic:  Temp: 20 C; (Comment: dynamic viscosity:von duennen
   Filmen.) (Ref. 120 handbook)
   Dynamic:  Temp: 121.7 - 273 C; (Ref. 121 handbook)
   Dynamic: 0.01779 g/(cm*sec); Temp: 30 C; (Ref. 122 handbook)
   Dynamic: 0.03825 - 0.563 g/(cm*sec); Temp:  -0.3 - -70.6 C;
   (Ref. 123 handbook)
   Dynamic: 2.33 - 1590 g/(cm*sec); Temp:  -96.5 - -135.5 C;
   (Ref. 124 handbook)
                                 -more-
?
       Display 6/CNTP/30

   Dynamic:  (Comment: dynamic viscosity:Einfluss von Drucken bis
   12000 kg/cmE2 auf die Viscositaet bei 30grad und 75grad.)
   (Ref. 125 handbook)
   Dynamic: 0.02237 - 0.00102 g/(cm*sec); Temp: 20 - 210 C; (Ref.
   126 handbook)
   Dynamic: 0.0229 g/(cm*sec); Temp: 20 C; (Ref. 54 handbook),
   (Ref. 127 handbook)
   Dynamic: 0.02015 g/(cm*sec); Temp: 25 C; (Ref. 54 handbook)
   Dynamic:  (Ref. 128 handbook),  (Ref. 129 handbook),  (Ref.
   130 handbook)
   Refs.
      54, Mumford, Phillips, JCSOA9, J.Chem.Soc., 1950 75, 81
      103, Timmermans, Delcourt, JCPBAN, J.Chim.Phys.Phys.Chim.
           Biol., 31(1934)104
```

```
114, Tonomura, STUCAV, Sci.Rep.Tohoku Univ.Ser.1, (1)22
     120, Zentralblatt: 1933 II 997
115, Mizushima, BCSJA8, Bull.Chem.Soc.Jpn., 1(1926)164


                         -more-


?


Display 6/CNTP/30


116, Trew, Watkins, TFSOA4, Trans.Faraday Soc., 29(1933)
     1311
117, Bridgman, Pr.am.Acad.Arts Sci., 77(1949)121,124, CA
     (1949)4067
118, Griengl, Kofler, Radda, MOCMB7, Monatsh.Chem., 62
     (1933) 139
119, Kobeko, Mitarb, ZFKHA9, Zh.Fiz.Khim., 9(1937)383, Acta
     physicoch.U.R.S.S., 6(1937)249, Zentralblatt: 1938 I
     3888
120, Rastow, Bowden, PRLAAZ, Proc.R.Soc.London A, 151(1935)
     226
121, Titani, BCSJA8, Bull.Chem.Soc.Jpn., 8(1933)267
122, Bridgman, Pr.am.Acad.Arts Sci., 61(1925/26),70,
     Zentralblatt: 1926 I,1919 II,1923
123, Mitsukuri, Tonomura, PIATA8, Proc.Imp.Acad.Tokyo,
     5,24, Zentralblatt: 1929 II,547
124, Tammann, Hesse, ZAACAB, Z.Anorg.Allg.Chem., 156
     (1926),250
125, Bridgman, Pr.am.Acad.Arts Sci., 61(1925/26),76,
     Zentralblatt: 1926 I,1919 II,1923
126, Lednewa, VMUNAE, Vestn.Mosk.Univ., 11 (1956) Nr.2, S.
     49, 56, CA (1956)11258
127, Yamakita, Fuyito, BICRAS, Bull.Inst.Chem.Res.Kyoto
     Univ., 24 (1951) 70


                         -more-


?


Display 6/CNTP/30

128, Thorpe, Rodger, Transact. Royal Soc.A, 185,533
129, Gartenmeister, ZEPCAC, Z.Phys.Chem.Stoechiom.
     Verwandtschaftsl., 6 (1890),529
130, Pribram, Handl, MOCMB7, Monatsh.Chem., 2 (1881),664

Kinematic:  (Comment: at:20-110 degreeC.) (Ref. 131 handbook)
Refs.
   131, Golik, et al, UKZHAU, Ukr.Khim.Zh.(Russ.Ed.), 21
        (1955) 167, 171
```

```
  Bulk:  (Comment: at:20-210 degreeC.) (Ref. 126 handbook)
  Refs.
     126, Lednewa, VMUNAE, Vestn.Mosk.Univ., 11 (1956) Nr.2, S.
          49, 56, CA (1956)11258

Self-Diffussion & Thermal Conductivity
  Self-diffusion:  Temp: 15 - 45 C; (Ref. 132 handbook)
  Refs.
     132, Partington, et al, JCPBAN, J.Chim.Phys.Phys.Chim.
          Biol., 55 (1958) 77

  Thermal Conductivity:  Temp: 20 - 90 C; (Ref. 133 handbook)

                          - end of display -
```

3N. SPECTRAL PROPERTIES

Spectroscopy has become a routine tool for the organic chemist in the past two or three decades. While the *Beilstein Handbook* does not contain complete spectral data —this is left to many specialized numeric data collection activities—there are limited spectral data and references to spectra in the current database. In general, as the first

TABLE XX

PROPERTY	SEARCH FIELD	DISPLAY FIELDS	
Electronic absorption (UV/VIS)	EAS	PP	SP
Electronic absorption (UV/VIS) max.	EAM	PP	SP
Fluorescence spectrum	FLS	PP	SP
Fluorescence maximum	FLM	PP	SP
Phosphorescence spectrum	PHS	PP	SP
Phosphorescence maximum	PHM	PP	SP
Infrared spectrum	IRS	PP	SP
Infrared maximum	IRM	PP	SP
Raman spectrum	RAMANS	PP	SP
Raman maximum	RAMANB	PP	SP
Mass spectrum, method	MSMETH	PP	SP
Mass spectrum, fragmentation	MSFRAG	PP	SP
Nuclear magnetic resonance spectrum	NMRS	PP	SP
Nuclear magnetic resonance absorption	NMRA	PP	SP
Electron spin resonance data	dp = RTESR	PP	SP
Nuclear quadrupole resonance	dp = RTNQR	PP	SP
Extinction absorption coefficient	dp = EAMEACC	PP	SP
Rotational spectrum	dp = RTROTS	PP	SP

segment of the Beilstein Database covers the literature up to 1959, there are less *nuclear magnetic resonance* (NMR) and *mass spectrometry* (MS) spectral data than other types. The types for which there are more data are *ultraviolet/visible*, which is called *electron absorption* (EA) in the Beilstein Database; *infrared* (IR), *fluorescence* (FL), and *phosphorescence* (PH). As the database is expanded, more spectral data will become available. The database format was designed as much for future data as for data now in the database. Hence there are empty fields now, but these "data gaps" will be filled in as time goes on.

In the current database, which has many spectral fields, the Electronic Absorption Maximum (EAM), the Fluorescence Maximum (FLM), the Infrared Maximum (IRM), and the Phosphorescence Maximum (PHM) contain numeric data. The Infrared Spectrum (IRS) field contains the frequency (wavenumber) range, and the Electronic Absorption Spectrum (EAS), the Fluorescence Spectrum (FLS), and the Phosphorescence Spectrum (PHS) contain the wavelength range over which the spectral data were reported in

TABLE XXI. Selected Solvents for UV Spectra in the Beilstein Factual Database

ACETANILIDE	ETHER	NACL
ACETONE	FECL3	NAHCO3
ACETONITRILE	FORMAMIDE	NAHPO3
ACRYLONITRILE	GLYCEROL	NAOH
ALCL3	H2O	NAPHTHALENE
ANILINE	H2S	NH3
BENZAMIDE	H2SO4	NITROBENZENE
BENZENE	H3PO4	OCTANE
BUTANE	HBR	PENTANE
CCL4	HCL	PENTANENITRILE
CH2CL2	HCLO4	PETROLEUM
CH2I2	HEPTANE	PHENOL
CHCL3	HEXANE	PHENYLMETHANOL
CHLOROBENZENE	HNO3	PIPERIDINE
CHLORONAPHTHALENE	KBR	PROPIONITRILE
CS2	KCL	PYRIDINE
CYCLOHEXANE	KCN	PYRROLIDINE
CYCLOHEXANOL	KH2PO4	QUINOLINE
CYCLOHEXANONE	KOH	STYRENE
D2O	LI2CO3	TETRACHLOROETHANE
DECAHYDRONAPHTHALENE	METHANOL	TETRAHYDROFURAN
DIACETOXYMETHANE	METHOXYBENZENE	TETRANITROMETHANE
DIMETHOXYMETHANE	METHYLCYCLOHEXANE	THIOPHENE
DIMETHYLFORMAMIDE	METHYLCYCLOPENTANE	TOLUENE
DIMETHYLSULFATE	NA2B4O7	TRICHLOROBENZENE
DIMETHYLSULFOXIDE	NA2CO3	TRIETHYLAMINE
DIOXANE	NA2HPO4	XYLENE
ETHANOL		

the literature. The **NMR** field only contains information on which nucleus was studied and what the solvent (if any) was used in obtaining the spectrum. No chemical shifts will be found in the current version of the database.

All the fields in the Beilstein Database which describe and contain spectral data are summarized in Table XX. These fields are a combination of numeric data and display only, and can be searched in the usual manner. The description of each parameter explains what is in the database, what can be searched, and what can only be displayed.

3N1. Electronic Absorption (UV) Spectrum

The UV, or *electronic absorption*, spectrum field (EAS) contains the range of absorption of the compound in the UV wavelength range of about 180–400 nm, along with solvent information, and the reference. A list of selected solvents is given in Table XXI.

Searches for the UV/VIS range and the solvent used in measuring the UV/VIS spectrum can be done alone or in conjunction with a number of other fields such as molecular weight or chemical elements. The example below is a search for compounds in the database for which there are UV/VIS absorption data in the range 400–420 nm, in which the solvent used was ethanol, and which have a molecular weight between 100 and 120. This search retrieves five records and the full basic information (BI) and spectroscopic data record (SP) for the first of the five is displayed, using the TYPE command:

```
?s eas=400:420(s)ethanol and mw=100:120
         2929   EAS=400 : EAS=420
       298006   ETHANOL  (see also A, aethanol, C2H5OH, EtOH)
         1363   (EAS)(S)ETHANOL
        23476   MW=100 : MW=120
   S1        5   EAS=400:420(S)ETHANOL AND MW=100:120

?t /bi gs sp/1

 1/BIGSSP/1
605871
cyano-acetic acid ethyl ester
  German Chem. Name: Cyan-essigsaeure-aethylester
  Lawson No: 1524, 298
  Beilstein Cit: 3-02-00-01628; 0-02-00-00585; 2-02-00-00531;
  1-02-00-00254; 4-02-00-01889; 5-02
  Molecular Formula: C5H7NO2
  Molecular Weight: 113.12
  No. of Ref: 515
```

Graphic Structure:
'605871'

Data Present:
```
 Data  Ref
 +Ref/Only UDF    Data Type
   16/1    PR Preparative Data
  245/307  CR Chemical Reactions
   99/28   PP Physical Properties
     /4    KW Short File Keywords
```
Spectra Data
   Nuclear Quadrupole Resonance
      Nuclear quadrupole resonance (Ref. 254)
      Refs.
         254, Onda, et al, JOMRA4, J.Magn.Reson., 18(1975)282
Vibrational Spectra
   IR Spectrum
      Wavenumber: 3030 - 833 cm**-1; Solvent: CCl4; (Ref. 33
      handbook)
      (Ref. 255)
      (Ref. 256)
      (Ref. 251)
      Refs.
         33, Rasmussen, Brattain, JACSAT, J.Amer.Chem.Soc.,
           71(1949)1074
         251, Laato, et al, ACSAA4, Acta Chem.Scand., 21(1967)
           2119, 2123,2129
         255, Charles, et al, JCFTBS, J.Chem.Soc.Faraday Trans.2,
           69(1973)1454,1461
         256, Inukai, KGKZA7, Kogyo Kagaku Zasshi, 70(1967)2395,
           2396
   Vibrational Spectrum
      Intensity of IR bands (Ref. 246 handbook), (Ref. 257
      handbook), (Ref. 258 handbook), (Ref. 259 handbook),
      (Ref. 260 handbook)
      Refs.
         246, Felton, Orr, JCSOA9, J.Chem.Soc., 1955 2170,2172
         257, Jesson, Thompson, SPACA5, Spectrochim.Acta,
           13(1959)217,219
         258, Skinner, Thompson, JCSOA9, J.Chem.Soc., 1955 487
         259, Thompson, Jameson, SPACA5, Spectrochim.Acta,
           13(1959)236,238
         260, Brown, JACSAT, J.Amer.Chem.Soc., 80(1958)3513

```
     Raman-spectrum:  (Ref. 34 handbook)
     Refs.
         34, Dadieu, Kohlrausch, MOCMB7, Monatsh.Chem., 55(1930)62,
             78, CHBEAM, Chem.Ber., 63(1930)257
Electronic Spectra
   UV/VIS Spectrum
       Wavelength: 230 - 300 nm; (Comment: unverd..) (Ref. 35
       handbook)
       Wavelength: 230 - 300 nm; Solvent: ethanol; (Ref. 35
       handbook)
       Wavelength: 240 - 280 nm; Solvent: ethanol; (Ref. 36
       handbook)
       Wavelength: 260 - 400 nm; Solvent: sodium ethanolate,
       ethanol; (Ref. 36 handbook) (Ref. 261)
       Refs.
           35, Schurz, et al, ZPCLAH, Z.Phys.Chem.(Leipzig), (N.
               F.)21(1959)185,187
           36, Segers, Bruylants, Ind. chim. belge Sonderband 27.
               Congr. int. Chim. ind. Bruessel 1954, Bd.3, S. 647
           261, Turkewitsch, Kowalin, UKZHAU, Ukr.Khim.Zh.
               (Russ.Ed.), 31(1965)607,609, CA 63(1965)7000
   Absorption Maxima
     (Ref. 262)
     Refs.
         262, Dale, Morgenlie, ACSAA4, Acta Chem.Scand., 24(1970)
             2408,2412
```

3N2. Electronic Absorption (UV/VIS) Spectrum Maximum

The UV/VIS, or electronic absorption, spectrum maximum field (EAM) contains up to six specific band absorptions shown by a compound in the UV/VIS spectral range (ca. 140–890 nm), along with solvent information and the reference. It is possible with this field to search, for example, for all compounds possessing a UV maximum at 340 nm when measured in methanol:

```
?s eam=340(s)methanol
           238   EAM=340
        138936   METHANOL  (see also CH3OH, Me, MeOH)
    S2      13   EAM=340(S)METHANOL

?t/k/4-7

 2/K/4
Electronic Spectra
   Absorption Maxima
       ...Wavelength: 340, 360 nm; Solvent: methanol; (Ref. 5
          handbook)
```

```
 2/K/5
Electronic Spectra
   Absorption Maxima
     Wavelength: 340 nm; Solvent: methanol; (Ref. 1 handbook)


 2/K/6
Electronic Spectra
   Absorption Maxima
     ...Wavelength: 340, 345 nm; Solvent: methanol; (Ref. 3
        handbook)

 2/K/7
Electronic Spectra
   Absorption Maxima
     Wavelength: 340 nm; Solvent: methanol; (Ref. 1 handbook)
```

If more than one peak is present in the UV spectrum, this may be enough to identify the compound:

```
?s eam=(340 and 360)(s)methanol
           238   EAM=340
           196   EAM=360
        138936   METHANOL  (see also CH3OH, Me, MeOH)
    S3       1   EAM=(340 AND 360)(S)METHANOL


?t/bi gs sp

 3/BIGSSP/1
273054
1,8-dimethoxy-naphthalene-2,3-dicarboxylic acid-anhydride
  German Chem. Name:
1,8-Dimethoxy-naphthalin-2,3-dicarbonsaeure-anhydrid
  Lawson No: 19230, 289
  Beilstein Cit: 4-18-00-02598; 5-18
  Molecular Formula: C14H10O5
  Molecular Weight: 258.23
  No. of Ref: 6
  Graphic Structure:
  '273054'
```

```
   Data Present:
     Data  Ref
     +Ref/Only UDF    Data Type
        2/1     PR Preparative Data
        1/1     CR Chemical Reactions
        6/      PP Physical Properties
 Electronic Spectra
    Absorption Maxima
        Wavelength: 247, 284, 300, 310 nm; Solvent: methanol; (Ref.
        5 handbook)
        Wavelength: 340, 360 nm; Solvent: methanol; (Ref. 5
        handbook)
        Refs.
           5, Nikuni, Hitsamoto, NNKKAA, Nippon Nogei Kagaku
              Kaishi, 20 (1944) 283, 287, CA (1951)5141
```

3N3. Fluorescence Spectrum

Subsequent to absorption, many compounds emit light from excited energy states created by the absorption of energy. If the emission is from an excited singlet state, it is called *fluorescence*. The fluorescence spectrum field contains the range of emission of the compound in the UV–visible range of about 350–700 nm, along with information as to the solvent, and the appropriate literature reference. Selected solvents are listed in Table XXII. The example below shows a search for compounds in the database, dissolved in ethanol, having fluorescent emission in the range of 400–420 nm:

```
?s fls=350:450(s)ethanol
          196    FLS=350 : FLS=450
        298006   ETHANOL  (see also A, aethanol, C2H5OH, EtOH)
    S4     50    FLS=350:450(S)ETHANOL

?d /cn k/14,15
        Display 4/CNK/14

3-hydroxy-2-(4-methoxy-phenyl)-chromen-4-one
Emission Spectra
    Fluorescence Spectrum
      Wavelength: 400 - 620 nm; Solvent: ethanol; (Ref. 19
      handbook)

                              - end of record -
```

```
?
        Display 4/CNK/15

1,3,6,8-tetramethyl-1 H,8 H-pteridine-2,4,7-trione
Emission Spectra
    Fluorescence Spectrum
      ...Wavelength: 350 - 530 nm; Solvent: ethanol; (Ref. 3
         handbook...

                               - end of display -
```

TABLE XXII.   Selected Solvents for Fluorescence Spectra in the Beilstein Factual Database

ACETONE	ETHANOL	NA2CO3
AQ	ETHER	NAOH
BENZENE	H2O	PENTANE
CCL4	H2SO4	PETROLEUM
CHCL3	HCL	POLYSTYRENE
CS2	HEPTANE	PYRIDINE
CYCLOHEXANE	HEXANE	TOLUENE
DIOXANE	METHANOL	XYLENE

3N4. Fluorescence Maximum

The fluorescence maximum field (FLM) contains up to five peaks for the maximum emission of the compound in the wavelength range of about 350–700 nm, along with solvent information and the literature reference. In the example below, the search is for compounds in the database having a fluorescence maximum at 395 nm in which water was the solvent used:

```
?s flm=395(s)(water or h2o)
             7   FLM=395
         31566   WATER  (see also H2O, w, wasser, wassers)
         82025   H2O  (see also w, wasser, wassers, water)
     S5      4   FLM=395(S)(WATER OR H2O)

?t /cn k/1-2

   5/CNK/1
adenosine
Emission Spectra
    Fluorescence Maxima
      Maxima: 395 nm; Solvent: H2O (pH 1); (Ref. 186 handbook)
```

```
  5/CNK/2
ATP
Emission Spectra
   Fluorescence Maxima
      Maxima: 395 nm; Solvent: H2O; (Comment: pH 1.) (Ref. 108
      handbook)
```

3N5. Phosphorescence Spectrum

Subsequent to absorption, many compounds emit light from excited energy states created by the absorption of energy. If the emission is from an excited triplet state, it is called *phosphorescence*. Phosphorescence occurs on a longer time scale than fluorescence because relatively long-lived excited states are involved. The phosphorescence spectrum field contains the emission of the compound in the wavelength range of about 400–700 nm, along with solvent information and the reference. A list of selected solvents is given in Table XXIII. The example below is a search for compounds in the database which have a phosphorescence emission in the range of 650–725 nm:

```
?s phs=650:725
      S6      2  PHS=650:725

?d /cn k/1
      Display 6/CNK/1

phenazine
Emission Spectra
   Phosphorescence Spectrum
      Wavelength: 650 - 725 nm; (Ref. 159 handbook),  (Ref. 164
      handbook

                          - end of record -
```

TABLE XXIII. Selected Solvents for Phosphorescence Spectra
in the Beilstein Factual Database

ACETANILIDE
BENZAMIDE
BUTANE
ETHANOL
ETHER

The identity of the solvent used in measuring the spectrum may also be specified in the search statement. Note that no qualifier is used with the name of the solvent, which must, however, be linked to the search statement with an (S) operator:

```
?s phs=350:750(s)methanol
            12   PHS=350 : PHS=750
        138936   METHANOL  (see also CH3OH, Me, MeOH)
     S7      4   PHS=350:750(S)METHANOL

?d /cn k/2
        Display 7/CNK/2

(+)-lysergic acid diethylamide
Emission Spectra
   Phosphorescence Spectrum
     Wavelength: 350 - 750 nm; Solvent: aq. methanol; (Comment:
     bei 77 K.) (Ref. 12 handbook)

                          - end of record -
```

3N6. Phosphorescence Maximum

The phosphorescence maximum field contains up to five peaks for the maximum emission of the compound in the wavelength range of about 350–700 nm, along with solvent information and the reference. The example below is a search for compounds in the database which show a phosphorescence maximum at 500 nm when measured in diethyl ether:

```
?s phm=500(s)diethyl ether
            1   PHM=500
        65180   DIETHYL ETHER   (see also Ae, aether,
                                     diaethylaether)
     S8     1   PHM=500(S)DIETHYL ETHER

?t /cn k

 8/CNK/1
2,5-dimethyl-pyrazine
Emission Spectra
   Phosphorescence Maxima
     Maxima: 363.63, 500 nm; Solvent: diethyl ether, 2-methyl-
     butane; (Comment: Loesungsm. (3): A; bei 77 K.) (Ref. 60...
```

TABLE XXIV. Selected Solvents for IR Spectra in the Beilstein Factual Database

ACETONE	DIOXANE	PARAFFIN
ACETONITRILE	ETHANOL	PENTANE
BENZENE	H2O	PERFLUOROKEROSENE
CCL4	HEXANE	PYRIDINE
CH2BR2	KBR	PYRROLE
CH2CL2	KCL	PYRROLIDINE
CHBR3	KI	SBCL3
CHCL3	METHANOL	TETRACHLOROETHANE
CS2	NACL	TETRAHYDROFURAN
CYCLOHEXANE	NEAT	THIOPHENE
D2O	NUJOL	VASELINE
DIMETHYLSULFOXIDE		

3N7. Infrared Spectrum

The *infrared* (IR) *spectrum* of a compound results from the absorption of energy in the IR wavelength region of 500–5000 cm^{-1} (wavenumbers). The IRS spectrum field in the Beilstein Database contains the range covered by the reported IR spectrum of the compound, along with solvent information and the reference. A list of selected solvents is given in Table XXIV. The example below is a search for compounds in the database which have an IR absorption between 1700 and 1800 cm^{-1} and in which the compound was dispersed in Nujol:

```
?s irs=1700:1800(s)nujol
            762   IRS=1700 : IRS=1800
           3247   NUJOL
      S9     94   IRS=1700:1800(S)NUJOL

?t /cn/3-9

 9/CN/3
eicosanoic acid

 9/CN/4
ethoxy-succinic acid diamide

 9/CN/5
ethoxy-butenedioic acid diamide

 9/CN/6
4,6-dioxo-hept-2 t-enedioic acid
```

```
 9/CN/7
DL-methionine S,S-dioxide

 9/CN/8
methionine S-imide-S-oxide

 9/CN/9
N-acetyl-DL-isoleucine
```

From this group of 94 records, two can be isolated as dealing with diamides:

```
?s diamide/cn and irs=1700:1800(s)nujol
            904   DIAMIDE/CN   (see also diamid)
            762   IRS=1700 : IRS=1800
           3247   NUJOL
             94   (IRS)(S)NUJOL
     S10       2   DIAMIDE/CN AND IRS=1700:1800(S)NUJOL

?t /cn k/1-2

 10/CNK/1
ethoxy-succinic acid diamide
ethoxy-succinic acid  diamide
Vibrational Spectra
   IR Spectrum
     Wavenumber: 10000 - 667 cm**-1; Solvent: nujol; (Ref. 2
     handbook)

 10/CNK/2
ethoxy-butenedioic acid diamide
ethoxy-butenedioic acid  diamide
Vibrational Spectra
   IR Spectrum
     Wavenumber: 10000 - 667 cm**-1; Solvent: nujol; (Ref. 3
     handbook)
```

3N8. Infrared Spectrum Maximum

The infrared maximum field (IRM) contains infrared absorption maxima as they are reported in the literature. A search of this field for a specific frequency will retrieve all compounds reported to have an absorption maximum at that frequency:

```
?s irm=1720
     S11      28   IRM=1720
```

```
?t 11/k/1-2

 11/K/1
Vibrational Spectra
   IR Bands
     Wavenumber: 1720 - 1560 cm**-1; (Ref. 5 handbook)

 11/K/2
Vibrational Spectra
   IR Bands
     Wavenumber: 1720 - 760 cm**-1; (Ref. 2 handbook)
```

These entries deal with specific compounds which may be of interest:

```
?t 11/cn sp/1

 11/CNSP/1
succinomonohydroxamic acid
Vibrational Spectra
   IR Bands
     Wavenumber: 1720 - 1560 cm**-1; (Ref. 5 handbook)
     Refs.
         5, Ames, Grey, JCSOA9, J.Chem.Soc., 1955 631,632
```

Ranges of frequencies may be used in this search:

```
?s irm=2000:3000
       S12    3602   IRM=2000:3000

?t /k/560

 12/K/560
Vibrational Spectra
   IR Bands
     Wavenumber: 3333 - 714 cm**-1; Solvent: KBr; (Ref. 6
     handbook...
```

and useful searches can result if this term is combined with a partial chemical name as well as the medium used in the infrared spectrum:

```
?s amine/cn and irm=2000:3000(s)kbr
          27986   AMINE/CN  (see also amin)
           3602   IRM=2000 : IRM=3000
           2812   KBR  (see also 3 related terms)
            846   (IRM)(S)KBR
    S13      82   AMINE/CN AND IRM=2000:3000(S)KBR

?t /cn gs k/30-33

 13/CNGSK/30

2-ethyl-6-methyl-2,3-dihydro-1 H-pyrrolo(3,4-c)pyridin-7-ylamine
2-ethyl-6-methyl-2,3-dihydro-1 H-pyrrolo(3,4-c)pyridin-7- ylamine
  Graphic Structure:
  '143569'
```

```
Vibrational Spectra
   IR Bands
     Wavenumber: 3030 - 2703 cm**-1; Solvent: KBr; (Comment:
     CH-Valenzschwingugngsbanden.) (Ref. 1 handbook)

 13/CNGSK/31
(4-methyl-thiazol-2-yl)-nitro-amine
4-methyl-thiazol-2-yl)-nitro- amine
  Graphic Structure:
  '140610'
```

```
Vibrational Spectra
   IR Bands
     Wavenumber: 3100 - 650 cm**-1; Solvent: KBr; (Ref. 3
     handbook)

 13/CNGSK/32
2,6-dimethyl-2,3-dihydro-1 H-pyrrolo(3,4-c)pyridin-7-ylamine
2,6-dimethyl-2,3-dihydro-1 H-pyrrolo(3,4-c)pyridin-7- ylamine
```

```
Graphic Structure:
 '139128'
```

```
Vibrational Spectra
   IR Bands
      Wavenumber: 3030 - 2703 cm**-1; Solvent: KBr; (Comment:
      CH-Valenzschwingungsbanden.) (Ref. 2 handbook)

 13/CNGSK/33
4-methyl-5-phenyl-(3)pyridylamine
4-methyl-5-phenyl-(3) pyridylamine
   Graphic Structure:
   '133807'
```

```
Vibrational Spectra
   IR Bands
      Wavenumber: 3448 - 1538 cm**-1; Solvent: KBr; (Ref. 1
      handbook)
```

3N9. Raman Spectrum

The *Raman spectrum* of a compound results from the absorption of energy in the IR wavelength region of 500–5000 cm^{-1} (wavenumbers). The Raman Spectrum field (**RAMANS**) in the Beilstein Database contains only the reference to the compound, along with some notes about the spectrum. The field can only be displayed, not searched. The example below is a search for compounds in the database which have a Raman field containing displayable information:

```
?s dp=ramans
      S14     561  DP=RAMANS  (Raman-spectrum)

?t /cn k/284-285


 14/CNK/284
but-2 t-ene
Vibrational Spectra
   Raman-spectrum:  Gas und Fluessigkeit (Ref. 260 handbook...
```

```
 14/CNK/285
3,4-dimethyl-hexane
Vibrational Spectra
   Raman-spectrum:  Ref. 73 handbook),  (Ref. 74 handbook),
   (Ref. 75 handbook),  (Ref...
```

3N10. Raman Spectrum Maximum

The Raman Bands (Raman spectrum maximum) field (RAMANB) in the Beilstein Database
contains records of compounds for which Raman spectrum maxima have been reported. This field is only searchable with the dp=ramanb search statement:

```
?s dp=ramanb
      S15      561  DP=RAMANB  (Raman-bands)

?d/cn k/105-106


      Display 15/CNK/105

penta-Si-chloro-Si-methyl-Si, Si'-ethanediyl-bis-silane
Vibrational Spectra
   Raman-bands: 3000-100cmE-1 (Ref. 1 handbook)

                              - end of record -

?
      Display 15/CNK/106

hexa-Si-ethyl-Si, Si'-ethanediyl-bis-silane
Vibrational Spectra
   Raman-bands: 3000-200 cmE-1 (Ref. 12 handbook)

                              - end of record -
```

3N11. Mass Spectrum

Two fields in the Beilstein Database are concerned with mass spectra. The MSMETH
field contains the mass spectral method that was reported, and the MSFRAG field carries
an indication that fragmentation of the compound is described in the original report. In
either case, the only information available is the reference to the spectrum. There are
no peaks or m/z values in the database. These fields can be used to retrieve mass
spectral data for different compounds. As an example, in the search below, a series of

esters of dicarboxylic acids for which mass spectra have been reported was retrieved:

```
?s mf=c11h20o4 and dp=msmeth
            446  MF=C11H2004
            416  DP=MSMETH  (Mass Spectrum Method)
     S16       6  MF=C11H2004 AND DP=MSMETH

?t 16/cn gs k/1-6

 16/CNGSK/1
2-methyl-adipic acid diethyl ester
  Molecular Formula:  C11H2004
  Graphic Structure:
  '1785459'
```

```
Other Spectra
   Mass Spectrum
      Method : Elektronenstoss; (Ref. 11 handbook)

 16/CNGSK/2
propyl-succinic acid diethyl ester
  Molecular Formula:  C11H2004
  Graphic Structure:
  '1782389'
```

```
Other Spectra
   Mass Spectrum
      Method : Elektronenstoss; (Ref. 5 handbook...

 16/CNGSK/3
4,4-dimethyl-heptanedioic acid dimethyl ester
  Molecular Formula:  C11H2004
  Graphic Structure:
  '1781853'
```

```
Other Spectra
    Mass Spectrum
        Method : Elektronenstoss; (Ref. 3 handbook)

  16/CNGSK/4
 2-ethyl-glutaric acid diethyl ester
   Molecular Formula:  C11H2004
   Graphic Structure:
   '1710288'
```

```
Other Spectra
    Mass Spectrum
        Method : Elektronenstoss; (Ref. 4 handbook)

  16/CNGSK/5
 heptanedioic acid diethyl ester
   Molecular Formula:  C11H2004
   Graphic Structure:
   '1710126'
```

```
Other Spectra
    Mass Spectrum
        Method : Elektronenstoss; (Ref. 31 handbook)

  16/CNGSK/6
 nonanedioic acid dimethyl ester
   Molecular Formula:  C11H2004
   Graphic Structure:
   '1710125'
```

```
Other Spectra
    Mass Spectrum
        Method : Elektronenstoss; (Ref. 29 handbook)
```

The MSFRAG field can be used to retrieve references to papers in which the mass spectral fragmentation of specific compounds is described:

```
?s diazepine/cn and dp=msfrag
           1197  DIAZEPINE/CN  (see also diazepin)
             47  DP=MSFRAG  (Mass Spectrum Fragmentation)
     S17       1  DIAZEPINE/CN AND DP=MSFRAG


?t/bi gs sp

 17/BIGSSP/1
106007
2-methyl-4,5,6,7-tetrahydro-1 H-(1,3)diazepine
  German Chem. Name: 2-Methyl-4,5,6,7-tetrahydro-1 H-(1,3)
  diazepin
  Lawson No: 28016
  Beilstein Cit: 4-23-00-00471; 5-23
  Molecular Formula: C6H12N2
  Molecular Weight: 112.17
  No. of Ref: 7
  Graphic Structure:
  '106007'
```

```
  Data Present:
   Data  Ref
   +Ref/Only UDF    Data Type
     1/       PR Preparative Data
     6/       PP Physical Properties
      /1      KW Short File Keywords
Spectra Data
  NMR Absorption
     Nucleus: 1H; Solvent: CDCl3; (Ref. 3 handbook)
     Refs.
        3, Desmarchelier, et al, AJCHAS, Aust.J.Chem., 21 (1968)
           257, 259
Vibrational Spectra
  IR Bands
     Wavenumber: 3300 - 1550 cm**-1; Solvent: neat (no solvent);
     (Ref. 4 handbook)
     Refs.
        4, Desmarchelier, et al, AJCHAS, Aust.J.Chem., 21 (1968)
           257, 258
```

```
Other Spectra
   Mass Spectrum
      Fragmentation; (Ref. 5 handbook)
      Refs.
         5, Desmarchelier, et al, AJCHAS, Aust.J.Chem., 21 (1968)
            257, 262
```

3N12. Nuclear Magnetic Resonance Spectrum

The *nuclear magnetic resonance* (NMR) spectrum of a compound comes from the fact
that many of its nuclei have spin and thus, when placed in a strong magnetic field, can
absorb energy of the proper frequency. To show NMR absorption a nucleus must not
have both an even atomic number and an even mass number. Thus ^{12}C and ^{16}O give no
NMR signal.

   The NMR spectral absorption field (NMRANUC) in the Beilstein Database contains
only the type of nucleus studied, the solvent, and the reference. In the example below a
search is carried out for any triazines whose ^{13}C NMR spectra have been reported:

```
?s triazine/cn and nmranuc=13c
            2233   TRIAZINE/CN  (see also triazin)
             193   NMRANUC=13C
     S18       1   TRIAZINE/CN AND NMRANUC=13C

?t /bi sp

 18/BISP/1
340805
2,4,6-tris-(1-phenyl-ethyl)-hexahydro-(1,3,5)triazine
   German Chem. Name:
2,4,6-Tris-(1-phenyl-aethyl)-hexahydro-(1,3,5)triazin
   Lawson No: 30044
   Beilstein Cit: 4-26-00-00276
   Molecular Formula: C27H33N3
   Molecular Weight: 399.58
   No. of Ref: 5
   Data Present:
    Data  Ref
    +Ref/Only UDF    Data Type
      11/2    PP Physical Properties
Spectra Data
   NMR Absorption
      Nucleus: 1H; Solvent: CDCl3; (Ref. 4 handbook)
      Nucleus: 1H; Solvent: benzene-d6; (Ref. 4 handbook)
      Nucleus: 13C; Solvent: CDCl3; (Ref. 4 handbook)
```

```
        Refs.
           4, Nielsen, et al, JOCEAH, J.Org.Chem., 39(1974)1349,
              1350,1353
Vibrational Spectra
   IR Bands
      Wavenumber: 3333 - 1370 cm**-1; Solvent: CHCl3; (Ref. 2
      handbook), (Ref. 4 handbook)
      Wavenumber: 3333 - 1370 cm**-1; Solvent: CS2; (Ref. 2
      handbook), (Ref. 4 handbook)
      Refs.
         2, Witkop, JACSAT, J.Amer.Chem.Soc., 78(1956)2873,2880
         4, Nielsen, et al, JOCEAH, J.Org.Chem., 39(1974)1349,
            1350,1353
```

3N13. Nuclear Magnetic Resonance Absorption

Nuclear magnetic resonance absorptions are included in the Beilstein Database and
stored in the **NMRANUC** field. This field can be used to find all NMR spectra recorded
using deuterochloroform or deuterated water as solvent:

```
?s dp=nmranuc(f)cdcl3
          1586   DP=NMRANUC   (Nuclear Magnetic Resonance Abs.
                               Nucleus)
           609   CDCL3  (see also 2 related terms)
     S19   386   DP=NMRANUC(F)CDCL3


?s dp=nmranuc(f)d2o
          1586   DP=NMRANUC   (Nuclear Magnetic Resonance Abs.
                               Nucleus)
           328   D20  (see also deuterium oxide, heavy water)
     S20    57   DP=NMRANUC(F)D2O
```

This sort of search can be combined with a search for specific types of compounds as
in the example below, in which purine diones whose NMR spectra have been recorded
in D_2O are retrieved:

```
?s (purine and dione)/cn and dp=nmranuc(f)d2o
          1744   PURINE/CN  (see also purin)
         24019   DIONE/CN  (see also dion)
          1586   DP=NMRANUC   (Nuclear Magnetic Resonance Abs.
                               Nucleus)
           328   D20  (see also deuterium oxide, heavy water)
            57   DP=NMRANUC(F)D2O
     S21    10   (PURINE AND DIONE)/CN AND DP=NMRANUC(F)D2O
```

```
?t /cn k/4-7

 21/CNK/4
1-methyl-3,7-dihydro-purine-2,6-dione
1-methyl-3,7-dihydro- purine -2,6- dione
Spectra Data
   NMR Absorption
      Nucleus: 1H; Solvent: dimethylsulfoxide-d6, D2O;
      (Comment: Spektrum,:des Kations, des Neutralmolekuels, des
       Monoanions und des Dianions bei 70grad:.) (Ref. 16...

 21/CNK/5
7-methyl-3,7-dihydro-purine-2,6-dione
7-methyl-3,7-dihydro- purine -2,6- dione
Spectra Data
   NMR Absorption
      Nucleus: 1H; Solvent: dimethylsulfoxide-d6, D2O;
      (Comment: Spektrum,:des Kations, des Neutralmolekuels, des
       Monoanions bei 70grad:.) (Ref. 35 handbook)

 21/CNK/6
9-methyl-3,9-dihydro-purine-2,6-dione
9-methyl-3,9-dihydro- purine -2,6- dione
Spectra Data
   NMR Absorption
      Nucleus: 1H; Solvent: dimethylsulfoxide-d6, D2O; (Comment:
      Spektrum,:des Kations, des Neutralmolekuels, des Monoanions
      und des Dianions bei 70grad:.) (Ref. 11...

 21/CNK/7
3,7-dimethyl-3,7-dihydro-purine-2,6-dione, theobromine
3,7-dimethyl-3,7-dihydro- purine -2,6- dione , theobromine
Spectra Data
   NMR Absorption
      Nucleus: 1H; Solvent: D2O; (Comment: Spektrun,:des
      Neutralmolekuels, des Anions bei 70grad:.) (Ref. 66,
handbook...
      Nucleus: 1H; Solvent: dimethylsulfoxide-d6, D2O; (Comment:
      Spektrum,:des Kations, des Neutralmolekuels, des Anions bei
      70grad:.) (Ref. 66 handbook...
```

The NMR Spectral Nucleus field (**NMRSNUC**) contains the type of nucleus whose absorption is reported as well as the solvent and a reference:

```
?s dp=nmrsnuc
    S22      567  DP=NMRSNUC  (Nuclear Magnetic Resonance Sp.
                              Nucleus)
```

This field can be used to find diethyl esters of malonic acid whose ^{13}C NMR spectra have been reported:

```
?s (malonic and diethyl(w)ester)/cn and nmrsnuc=13c
           3908   MALONIC/CN
          15348   DIETHYL/CN  (see also diaethy, diaethyl)
          72717   ESTER/CN
           7992   DIETHYL/CN(W)ESTER/CN
             51   NMRSNUC=13C
    S23      1   (MALONIC AND DIETHYL(W)ESTER)/CN AND
                    NMRSNUC=13C

?t /bi gs sp

 23/BIGSSP/1
175587
tetrahydro(2)furylidene-malonic acid diethyl ester
  German Chem. Name:
Tetrahydro(2)furyliden-malonsaeure-diaethylester
  Lawson No: 19652, 298
  Beilstein Cit: 4-18-00-04454; 5-18
  Molecular Formula: C11H1605
  Molecular Weight: 228.24
  No. of Ref: 6
  Graphic Structure:
  '175587'
```

```
  Data Present:
   Data  Ref
   +Ref/Only UDF    Data Type
      4/4      PR Preparative Data
     10/       PP Physical Properties
      /1       KW Short File Keywords
Spectra Data
   NMR Spectrum
      Nucleus: 1H, 13C; Solvent: CCl4; (Ref. 4 handbook)
      Refs.
         4, Svendsen, Boll, ACSAA4, Acta Chem.Scand., (B) 29
            (1975) 197,199
Electronic Spectra
   Absorption Maxima
      Wavelength: 247 nm; Solvent: ethanol; (Ref. 3 handbook)
      Refs.
         3, Haynes, et al, JCSOA9, J.Chem.Soc., 1956 4661, 4664
```

3N14. Electron Spin Resonance Spectrum

If the electron spin resonance spectrum of a compound is reported, that fact is noted in the RTESR field. This field must be searched with the dp=rtesr search statement:

```
?s dp=rtesr
        S24    3885   DP=RTESR   (ESR Data)

?t /k/1000-1001

 24/K/1000
Spectra Data
   ESR Data
       ESR Spectrum (Ref. 1)

 24/K/1001
Spectra Data
   ESR Data
       ESR Spectrum (Ref. 2)
```

This field can be used to determine if the ESR spectral data for specific compounds have been published:

```
?s glycine/cn and dp=rtesr
            2218  GLYCINE/CN  (see also glycin)
            3885  DP=RTESR   (ESR Data)
      S25      9  GLYCINE/CN AND DP=RTESR

?t/k/1-3

 25/K/1
pentadeuterio-glycine
Spectra Data
   ESR Data
       ESR Spectrum (Ref. 4...

 25/K/2
N-glycyl-glycine
Spectra Data
   ESR Data
       ESR Spectrum (Ref. 105 handbook),  (Ref. 106 handbook),
       (Ref. 107 handbook...
```

```
 25/K/3
N-carbamimidoyl-glycine
Spectra Data
   ESR Data
       ESR Spectrum (Ref. 46 handbook)
```

3N15. Nuclear Quadrupole Resonance Spectrum

When the nuclear quadrupole resonance spectrum of a compound is reported, that fact is noted in the **RTNQR** field of the Beilstein Database. This field can only be searched with the dp=rtnqr search statement:

```
?s dp=rtnqr
      S26    2683  DP=RTNQR  (Nuclear Quadrupole Resonance)

?t /cn k/100-102

 26/CNK/100
Spectra Data
   Nuclear Quadrupole Resonance
       Nuclear quadrupole resonance (Ref. 10)

 26/CNK/101
Spectra Data
   Nuclear Quadrupole Resonance
       Nuclear quadrupole resonance (Ref. 3)

 26/CNK/102
Spectra Data
   Nuclear Quadrupole Resonance
       Nuclear quadrupole resonance (Ref. 1)
```

This field is useful when trying to locate records of compounds whose NQR data have been reported.

3N16. Rotational Spectra

If rotational spectra have been reported for a compound, the **RTROTS** field is modified to reflect this. The **RTROTS** field may only be searched with the dp=rtrots search statement:

```
?s dp=rtrots
     S28     2343   DP=RTROTS  (Rotational Spectrum)

?d /cn k/2200
        Display 28/CNK/2200

chlorocarbonic acid ethyl ester
Spectra Data
   Rotational Spectrum
       Microwave spectrum (Ref. 23)

                             - end of record -
```

Searches may be carried out using specific terms in connection with the **RTROTS** field:

```
?s dp=rtrots(f)stark effect
            2343   DP=RTROTS  (Rotational Spectrum)
            384   STARK EFFECT
     S29    346   DP=RTROTS(F)STARK EFFECT

?d /cn k/120-121
        Display 29/CNK/120

Spectra Data
   Rotational Spectrum
       Stark effect (Ref. 53),  (Ref. 54

                             - end of record -
?
        Display 29/CNK/121

Spectra Data
   Rotational Spectrum
       Stark effect (Ref. 61)

                             - end of display -
```

and information concerning specific effects shown by specific compounds can be obtained with normal Boolean logic:

```
?s octylamine/cn and dp=rtrots(s)stark effect
            204  OCTYLAMINE/CN  (see also octylamin)
           2343  DP=RTROTS  (Rotational Spectrum)
            384  STARK EFFECT
            347  DP=RTROTS(S)STARK EFFECT
     S30      2  OCTYLAMINE/CN AND DP=RTROTS(S)STARK EFFECT

?t/k/2

 30/K/
 octylamine
Spectra Data
   Rotational Spectrum
       Stark effect (Ref. 47)
```

3N17. Extinction/Absorption Coefficient Data

Data concerning extinction coefficients or absorption coefficients will be stored in the EAMEAC field of the Beilstein Database, but currently this field is empty:

```
?s dp=eameacc
     S31      0  DP=EAMEACC
```

3O. Controlled Terms

Published data on the various properties of chemical compounds in the Beilstein Database are often referenced but not included in the searchable fields of the database.

Data of this sort are stored in the Controlled Terms fields, whose mnemonics always begin with RT. There are 28 such RT*XXX* fields, ranging from affinity (RTAFF) to vibrational spectra (RTVIBSC). In addition, each of these is accompanied by a comments field, which has the same mnemonic, with a final C added, as in RTAFFC.

The data within these fields are not searchable, but the fields can be displayed and they can be retrieved with a DP= search qualifier:

```
?s dp=rtemol
      S1   1226  DP=RTEMOL  (Molecular Energy)

?t /k/1-3
```

```
 1/K/1
Molecular Energy Parameters
   Molecular Energy
      Bond  energy (Ref. 1)

 1/K/2
Molecular Energy Parameters
   Molecular Energy
      Bond  energy (Ref. 1)

 1/K/3
Molecular Energy Parameters
   Molecular Energy
      Bond  energy (Ref. 1)
```

The Controlled Terms fields that are currently in the database can all be found with an
EXPAND command:

```
?e dp=rt

Ref    Items    RT   Index-term
E1    262980         DP=RIW (Refractive Index Wavelength)
E2   1321113         DP=RN
E3         0        *DP=RT
E4       795     2   DP=RTAFF (Affinity)
E5       795         DP=RTAFFC (Affinity Comments/refs.)
E6     17471     1   DP=RTANAL (Elementary Analysis)
E7     17471         DP=RTANALC (Elementary Analysis Comments/refs.)
E8      5688     9   DP=RTCAL (Calorific Data)
E9      5688         DP=RTCALC (Calorific Data Comments/refs.)
E10      432     2   DP=RTCOMP (Compressibility)
E11      432         DP=RTCOMPC (Compressibility Comments/refs.)
E12    22636     3   DP=RTCONF (Conformation)

        Enter P or E for more


? p

Ref    Items    RT   Index-term
E13    22636         DP=RTCONFC (Conformation Comments/refs.)
E14    10662     9   DP=RTCP (Crystal Phase)
E15    10662         DP=RTCPC (Crystal Phase Comments/refs.)
E16    10789     3   DP=RTCPL (Coupling Constants)
E17    10789         DP=RTCPLC (Coupling Comments/refs.)
E18     3516    15   DP=RTEL (Electric Data)
E19     3516         DP=RTELC (Electric Data Comments/refs.)
```

```
E20     7525    18   DP=RTELCH (Electrochemical Behavior)
E21     7525         DP=RTELCHC (Electrochemical Behavior
Comments/refs.)
E22      469     2   DP=RTELPOL (Electrical Polarizability)
E23      469         DP=RTELPOLC (Electrical Polarizability
Comments/refs.)
E24     1441     7   DP=RTELS (Electronic Spectrum)

        Enter P or E for more

? p

Ref    Items   RT   Index-term
E25     1441         DP=RTELSC (Electronic Spectrum Comments/refs.)
E26     1226     2   DP=RTEMOL (Molecular Energy)
E27     1226         DP=RTEMOLC (Molecular Energy Comments/refs.)
E28     4127    11   DP=RTEMS (Emission Spectrum)
E29     4127         DP=RTEMSC (Emission Spectrum Comments/refs.)
E30     3885     6   DP=RTESR (ESR Data)
E31     3885         DP=RTESRC (ESR Data Comments/refs.)
E32      103     1   DP=RTGAS (Association In The Gas Phase)
E33      103         DP=RTGASC (Association In The Gas Phase
Comments/refs.)
E34     1314     8   DP=RTLIQ (Liquid Phase)
E35     1314         DP=RTLIQC (Liquid Phase Comments/refs.)
E36      512     2   DP=RTMAG (Magnetic Data)

        Enter P or E for more

? p

Ref    Items   RT   Index-term
E37      512         DP=RTMAGC (Magnetic Data Comments/refs.)
E38     3739     2   DP=RTMDEF (Molecular Deformation)
E39     3739         DP=RTMDEFC (Molecular Deformation
Comments/refs.)
E40      800     6   DP=RTMEC (Mechanical Properties)
E41      800         DP=RTMECC (Mechanical Properties
Comments/refs.)
E42     4088    11   DP=RTNMR (NMR Data)
E43     4088         DP=RTNMRC (NMR Data Comments/refs.)
E44     2683     1   DP=RTNQR (Nuclear Quadrupole Resonance)
E45     2683         DP=RTNQRC (Nuclear Quadrupole Resonance
Comments/refs.)
E46     4657    18   DP=RTOPT (Optics)
E47     4657         DP=RTOPTC (Optics Comments/refs.)
E48     5366     6   DP=RTOSM (Other Spectroscopic Methods)

        Enter P or E for more
```

```
? p

Ref     Items    RT   Index-term
E49     5366          DP=RTOSMC (Other Spectroscopic Methods
Comments/refs.)
E50     2343      5   DP=RTROTS (Rotational Spectrum)

? p

Ref     Items    RT   Index-term
E1      2343      5   DP=RTROTS (Rotational Spectrum)
E2      2343          DP=RTROTSC (Rotational Spectrum Comments/refs.)
E3   1149449     25   DP=RTSF (Short File Keyword)
E4   1149449          DP=RTSFC (Short File Keyword Comments/refs.)
E5      6651      2   DP=RTSKEL (Interatomic Distances And Angles)
E6      6651          DP=RTSKELC    (Interatomic    Distances    And
  Angles
                      Comments/refs)
E7       555      3   DP=RTUP (Ultrasonic Properties)
E8       555          DP=RTUPC (Ultrasonic Properties Comments/refs.)
E9      2250      9   DP=RTVIBS (Vibrational Spectrum)
E10     2250          DP=RTVIBSC (Vibrational Spectrum
Comments/refs.)
E11    77566          DP=RXNAIM (Chemical Reaction Aim Of The Study)
E12   377609          DP=RXNC (Chemical Reaction Comments/refs.)

         Enter P or E for more
```

IV. Displaying Data from the Beilstein Database

Once a compound of interest has been located in the Beilstein Database, it is necessary to be able to review the factual data associated with it. For many compounds, the quantity of data in the file is very large and much of it may be unrelated to the question at hand.

Because of this, it is useful to be able to display subsets of the factual data, and the system permits this by means of hierarchically organized display fields. In this section, a simple compound, 2-bromothiophene, will be used as an example, and various data displays will be developed to illustrate the organization of the display formats:

```
?s 2-bromo-thiophene/cn
     S1          1   2-BROMO-THIOPHENE/CN
```

This single compound will be referenced whenever results set S1 is invoked, and as long as S1 is the only results set, it need not be identified in any display commands.

4A. DISPLAY OF BASIC INFORMATION

The quickest way to learn the full identity of the chemical compound which is the subject of a record is to *display* or *type* it (d or t, respectively) using format 2 or the user-defined format (UDF) BI. A display in format 2 gives the identity of the compound and a complete list of information available. Format BI (below) gives identity data and a summary of the data available.

```
?t /bi

 1/BI/1
104663
2-bromo-thiophene
   German Chem. Name: 2-Brom-thiophen
   Lawson No: 16857
   Beilstein Cit: 4-17-00-00245; 0-17-00-00033; 2-17-00-00036;
                  1-17-00-00018; 5-17
   Molecular Formula: C4H3BrS
   Molecular Weight: 163.03
   No. of Ref: 130
   Data Present:
    Data  Ref
    +Ref/Only UDF    Data Type
      10/1       PR Preparative Data
      33/11      CR Chemical Reactions
      60/25      PP Physical Properties
       5/        DR Characterization Derivative
        /8       KW Short File Keywords
```

This provides the identification data for the compound (name, molecular formula, and so on) and also provides a very terse summary of the remainder of the record. In this case, we see that there are 130 references, 11 preparative methods (one of which is only cited as a reference), etc. This basic information display will usually tell you whether you wish to look at more of the record or not.

The chemical structure of the compound can be requested with this basic information, by adding gs to the format. Note, however, that this only produces the ROSDAL string (see section 5D1) for the compound. To print a structure diagram, you must use the auxiliary output program GEOFF, which is described in section 5E4:

```
?t/bi gs

 1/BIGS/1
104663
2-bromo-thiophene
   German Chem. Name: 2-Brom-thiophen
   Molecular Formula: C4H3BrS
   Molecular Weight: 163.03
   No. of Ref: 130
Graphic Structure:
'104663'
```

```
   Data Present:
    Data   Ref
    +Ref/Only UDF    Data Type
      10/1      PR Preparative Data
      33/11     CR Chemical Reactions
      60/25     PP Physical Properties
       5/       DR Characterization Derivative
       /8       KW Short File Keywords
```

The table at the end of this display indicates that for this compound, there are 10 preparative routes described in detail and one published route cited; 33 chemical reactions described, as well as another 11 for which literature citations are given; and so on. This summary material can be broken down further as described in the next section.

Use of format 12 will give a similar, but more detailed summary, which includes all the references:

```
?t /12

 1/12/1
104663
2-bromo-thiophene
   German Chem. Name: 2-Brom-thiophen
   Molecular Formula: C4H3BrS
   Molecular Weight: 163.03
   No. of Ref: 130
Graphic Structure:
'104663'
```

```
   Data Present:
    Data  Ref
    +Ref/Only UDF     Data Type
      10/1       PR Preparative Data
      10/1          .Preparation
      33/11      CR Chemical Reactions
      60/25      PP Physical Properties
       4/7       SE .Structure & Energy Parameters
        /2          ..Interatomic Distances & Angles
       3/            ..Dipole Moment
        /1           ..Optical Anisotropy
        /1           ..Coupling Phenomena
        /2           ..Molecular Deformation & Potential
       1/1           ..Molecular Energy Parameters
       1/            ...Dissociation Energy
        /1           ...Ionization Potential
      56/18         .Physical Properties of Pure Compound
      24/       PS ..Physical State
      24/       PL ...Liquids
      24/       BP ....Boiling Point
      10/4      PM ..Other Physical & Mechanical Properties
      10/       DN ...Density
        /4          ...Surface Tension
      16/       OP ..Optics
      16/       RI ...Refractive Index
       6/13     SP ..Spectra
       1/3          ...NMR
```

```
         1/2          ....NMR Spectrum
          /1          ....NMR Absorption
          /1          ...Nuclear Quadrupole Resonance
         3/4          ...Vibrational Spectra
         3/1      IR ....IR Spectrum
          /3          ....Raman Spectrum
         2/3          ...Electronic Spectra
         2/1          ....UV/VIS Spectrum
          /2          ....Absorption Maxima
          /2          ...Other Spectra
          /1          ....Mass Spectrum
          /1      EB ..Electrochemical Behavior
         5/       DR Characterization Derivative
          /8      KW Short File Keywords
Refs.
   1, V Meyer, CHBEAM, Chem.Ber., 16 (1883), 1469, 1472
   2, Toehl, Schultz, CHBEAM, Chem.Ber., 27 (1894), 2835
         .
         .
         .
   129, Gronowitz, Hoffman, ARKEAD, Ark.Kemi, 16(1961)539,542,543
   130, Watanabe, et al, JQSRAE, J.Quant.Spectrosc.Radiat.
   Transfer, 2(1962)369,379
```

4B. DETAIL OF PROPERTIES DATA

It is possible, using format 2, to obtain the basic information, followed by a breakdown of the physical properties data for the compound:

```
? t/2

 1/2/1
104663
2-bromo-thiophene
  German Chem. Name: 2-Brom-thiophen
  Molecular Formula: C4H3BrS
  Molecular Weight: 163.03
  No. of Ref: 130
  Data Present:
   Data  Ref
   +Ref/Only UDF    Data Type
     10/1     PR Preparative Data
     10/1        .Preparation
```

```
   33/11   CR Chemical Reactions
   60/25   PP Physical Properties
    4/7    SE .Structure & Energy Parameters
     /2       ..Interatomic Distances & Angles
    3/        ..Dipole Moment
     /1       ..Optical Anisotropy
     /1       ..Coupling Phenomena
     /2       ..Molecular Deformation & Potential
    1/1       ..Molecular Energy Parameters
    1/        ...Dissociation Energy
     /1       ...Ionization Potential
   56/18      .Physical Properties of Pure Compound
   24/    PS ..Physical State
   24/    PL ...Liquids
   24/    BP ....Boiling Point
   10/4   PM ..Other Physical & Mechanical Properties
   10/    DN ...Density
     /4       ...Surface Tension
   16/    OP ..Optics
   16/    RI ...Refractive Index
    6/13  SP ..Spectra
    1/3       ...NMR
    1/2       ....NMR Spectrum
     /1       ....NMR Absorption
     /1       ...Nuclear Quadrupole Resonance
    3/4       ...Vibrational Spectra
    3/1   IR ....IR Spectrum
     /3       ....Raman Spectrum
    2/3       ...Electronic Spectra
    2/1       ....UV/VIS Spectrum
     /2       ....Absorption Maxima
     /2       ...Other Spectra
     /1       ....Mass Spectrum
     /1    EB ..Electrochemical Behavior
    5/     DR Characterization Derivative
     /8    KW Short File Keywords
```

No further detail of the preparations or reactions data is given, but the 60/25 Physical Properties given in the BI format are now broken down in detail.

4C. PHYSICAL PROPERTIES DATA

If one wishes to examine a specific physical property, the subsection to which it belongs, identified as a UDF (user-defined format) in the output above, can be cited as a format. Thus in the above case, IR will provide the IR spectral data and SP will give the NMR, vibrational, and other spectral data. In both cases, the relevant references are also given:

```
?t/ir

 1/IR/1
Vibrational Spectra
   IR Spectrum
      Wavenumber: 5000 - 667 cm**-1; Solvent: 2-bromo-thiophene;
      (Ref. 57 handbook),  (Ref. 58 handbook)
      Wavenumber: 667 - 435 cm**-1; Solvent: 2-bromo-thiophene;
    (Ref. 59 handbook)
      Wavenumber: 2500 - 400 cm**-1; Solvent: 2-bromo-thiophene;
      (Ref. 60 handbook) (Ref. 35)
      Refs.
         35, Horak, et al, SPACA5, Spectrochim.Acta, 22(1966)
         1355,1361
         57, Gronowitz, ARKEAD, Ark.Kemi, 7(1954/55)267,269,271
         58, A.P.I.Res.Project, 44 Nr.966(1949)
         59, A.P.I.Res.Project, 44 Nr. 967(1949)
         60, Hidalgo, JPRAAJ, J.Phys.Radium, (8)16(1955)
         366,368,369

?t /sp

 1/SP/1
Spectra Data
   NMR Spectrum
      Nucleus: 1H, 13C; Solvent: cyclohexane; (Ref. 50 handbook),
      (Ref. 51 handbook) (Ref. 52) (Ref. 31)
      Refs.
         31, Hoffman, Gronowitz, ARKEAD, Ark.Kemi, 16(1961)515,
         519,520,523,531,532
         50, Gronowitz, Hoffman, ARKEAD, Ark.Kemi, 13(1958/59)
         279,281
         51, Takahashi, et al, BCSJA8, Bull.Chem.Soc.Jpn.,32
         (1959)156,159
         52, Dhingra, et al, PISAA7, Proc.-Indian Acad.Sci.
         Sect.A, 65(1967)203,205
   NMR Absorption
      (Ref. 53)
      Refs.
         53, Dhingra, et al, PISAA7, Proc.-Indian Acad.Sci.
         Sect.A,65(1967)203,205
   Nuclear Quadrupole Resonance
      Nuclear quadrupole resonance (Ref. 54), (Ref. 55), (Ref.56)
      Refs.
```

```
        54, Grechishkin, et al, TEXCAK, Theor.Exp.Chem.
        (Engl.Transl.), 7(1971)579
        55, Grechishkin, Shlykov, JSTCAM, J.Struct.Chem.
        (Engl.Transl.), 16(1975)127
        56, Dormidontov, et al, ORMRBD, Org.Magn.Reson.,
        4(1972)599,600
Vibrational Spectra
   IR Spectrum
      Wavenumber: 5000 - 667 cm**-1; Solvent: 2-bromo-thiophene;
      (Ref. 57 handbook), (Ref. 58 handbook)
      Wavenumber: 667 - 435 cm**-1; Solvent: 2-bromo-thiophene;
      (Ref. 59 handbook)
      Wavenumber: 2500 - 400 cm**-1; Solvent: 2-bromo-thiophene;
      (Ref. 60 handbook) (Ref. 35)
      Refs.
        35, Horak, et al, SPACA5, Spectrochim.Acta, 22(1966)
        1355,1361
        57, Gronowitz, ARKEAD, Ark.Kemi, 7(1954/55)267,269,271
        58, A.P.I.Res.Project, 44 Nr.966(1949)
        59, A.P.I.Res.Project, 44 Nr. 967(1949)
        60, Hidalgo, JPRAAJ, J.Phys.Radium, (8)16(1955)
        366,368,369
   Raman-spectrum: (Ref. 61 handbook)
      (Ref. 35)
      (Ref. 36)
   Refs.
      35, Horak, et al, SPACA5, Spectrochim.Acta, 22(1966)
      1355,1361
      36, Davidovics, et al, SAMCAS, Spectrochim.Acta Part A,
      23(1967)2669,2674
      61, Simon, Kirret, NATWAY, Naturwissenschaften, 28(1940)47
Electronic Spectra
UV/VIS Spectrum
      Wavelength: 210 - 285 nm; Solvent: hexane; (Ref. 62
      handbook)
      Wavelength: 215 - 280 nm; Solvent: ethanol; (Ref. 63
      handbook) (Ref. 12)
      Refs.
        12, Deganti, et al, BSFCAY, Boll.Sci.Fac.Chim.Ind.
            Bologna, 19(1961)76,77,81
        62, Pappalardo, GCITA9, Gazz.Chim.Ital., 89(1959)540,
            542,544
        63, Gronowitz, ARKEAD, Ark.Kemi, 13(1958/59)239,243
   Absorption Maxima
      (Ref. 42)
      (Ref. 47)
      Refs.
```

```
        42, Jeffery, et al, JCSOA9, J.Chem.Soc., (1961)570
        47, Tundo, BSFCAY, Boll.Sci.Fac.Chim.Ind.Bologna,
        18(1960)102,103
Other Spectra
   Other Spectroscopic Methods
      Photoelectron spectrum (Ref. 37),  (Ref. 64),  (Ref. 65)
      Refs.
         37, Rabalais, et al, IJMIBY, Int.J.Mass Spectrom.Ion
         Phys., 9(1972)185
         64, Baker, et al, ANCHAM, Anal.Chem., 42(1970)1064
         65, Bozic, et al, JCPKBH, J.Chem.Soc.Perkin Trans.2,
         (1977)1413
   Mass Spectrum
      (Ref. 64)
      Refs.
         64, Baker, et al, ANCHAM, Anal.Chem., 42(1970)1064
```

4D. PREPARATIONS DATA

The Beilstein Database contains 10 preparative methods described and 1 cited for 2-bromothiophene. These can be displayed or typed with a user-defined format of PR. This will list the preparative methods, together with the relevant literature references:

```
?t/pr

 1/PR/1
  Data Present:
   Data  Ref
   +Ref/Only UDF    Data Type
     10/1     PR Preparative Data
Preparative Data
   Preparation
      Starting Material: thiophene, bromine
      Other Conditions: Behandeln des Produkts mit Wasser;
      anschliessend mit Natronlauge und mit alkoh. Kali
      By-product: 2.5-dibromo-thiophene (Ref. 1 handbook)
   Preparation
      Starting Material: bromine, thiophene
      Reagent: glacial acetic acid (Ref. 2 handbook)
   Preparation
      Starting Material: 2.5-dibromo-thiophene
      Reagent: magnesium
      Other Conditions: Behandeln der Magnesiumverbindung mit
      Salzsaeure (Ref. 3 handbook)
```

```
Preparation
    Starting Material: thiophene, bromine
    Reagent: glacial acetic acid (Ref. 4 handbook)
Preparation
    Starting Material: thiophene, acetic acid bromoamide (Ref.
    5 handbook)
Preparation
    Starting Material: thiophene, cyanogen bromide
    Reagent: bromine
    Temp: 45 - 50 C
    By-product: 2.5-dibromo-thiophene (Ref. 6 handbook),  (Ref.
    7 handbook)
Preparation
    Starting Material: thiophene, dioxane-bromo-adduct
    Reagent: diethyl ether (Ref. 8 handbook)
Preparation
    Starting Material: bromo-vapour, thiophene
    Temp: 80 C (Ref. 9 handbook)
Preparation
    Starting Material: thiophene, bromine
    Reagent: tetrachloromethane
    By-product: 2,5-bromo-thiophene (Ref. 10 handbook)
Preparation
    Starting Material: 2,5-dibromo-thiophene
    Reagent: copper, quinoline (Ref. 11 handbook)
Preparation (Ref. 12), (Ref. 13), (Ref. 14), (Ref. 15), (Ref.
    16), (Ref. 17), (Ref. 18), (Ref. 19), (Ref. 20), (Ref. 21)
Refs.
    1, V Meyer, CHBEAM, Chem.Ber., 16 (1883), 1469, 1472
    2, Toehl, Schultz, CHBEAM, Chem.Ber., 27 (1894), 2835
    3, Gattermann, LACHDL, Liebigs Ann.Chem., 393 (1912), 230
    4, Krause, Renwanz, CHBEAM, Chem.Ber., 62 (1929), 1710,
    CHBEAM, Chem.Ber., 65 (1932), 778
    5, Steinkopf, Otto, LACHDL, Liebigs Ann.Chem., 424 (1921),
    69, 70
    6, Steinkopf, Koehler, LACHDL, Liebigs Ann.Chem., 52 (1936), 22
    7, Steinkopf, LACHDL, Liebigs Ann.Chem., 430 (1923), 99
    8, Terentew, et al, ZOKHA4, Zh.Obshch.Khim., 24(1954)
    1265,1269;engl.Ausg.S.1251,1254
    9, Michigan Chem Corp, US 2544164, (1947)
    10, Blicke, Burckhalter, JACSAT, J.Amer.Chem.Soc.,
    64(1942)477,478
    11, Nishimura, et al, Bl.Univ.Osaka Prefect.,
    (A)6(1958)127,128,130
    12, Deganti, et al, BSFCAY, Boll.Sci.Fac.Chim.Ind.Bologna,
    19(1961)76,77,81
    13, Nemec, et al, CCCCAK, Collect.Czech.Chem.Commun.
```

```
37(1972)3122
14, Pearson, et al, SYNTBF, Synthesis, (1976)621
15, Jones, Moodie, JSOOAX, J.Chem.Soc.C, (1960)2021
16, McKillop, et al, JOCEAH, J.Org.Chem., 37(1972)88
17, Chrzaszczewska, Szalecki, SLACBC, Soc.Sci.Lodz.Acta
Chim., 12(1967)119,122
18, Barker, et al, JCCCAT, J.Chem.Soc.Chem.Commun., (1972)615
19, Trompen, Huisman, RTCPA3, Recl.Trav.Chim.Pays-Bas,
85(1966)175,179
20, Olsson, Axen, ARKEAD, Ark.Kemi, 22(1964)237,240
21, van der Plas, Persoons, RTCPA3, Recl.Trav.Chim.Pays-Bas,
83(1964)701,704
```

A simple way to extract and display a single entry from this list is with a second search, followed by display of the KWIC format. As an example, to retrieve the sixth preparative method listed above, a search for the compound, with the correct starting material (cyanogen bromide) and reference (6), followed by a type command gives the desired result:

```
?s 2-bromo-thiophene/cn and start=cyanogen bromide and rf=6

              1   2-BROMO-THIOPHENE/CN
            323   START=CYANOGEN BROMIDE
          88592   RF=6
    S2        1   2-BROMO-THIOPHENE/CN AND START=CYANOGEN BROMIDE
                  AND RF=6


?t/bi gs k

 2/BIGSK/1
104663
2-bromo-thiophene
   German Chem. Name: 2-Brom-thiophen
   Lawson No: 16857
   Beilstein Cit: 4-17-00-00245; 0-17-00-00033; 2-17-00-00036;
                  1-17-00-00018; 5-17
   Molecular Formula: C4H3BrS
   Molecular Weight: 163.03
   No. of Ref: 130
Graphic Structure:
'104663'
```

```
   Data Present:
    Data  Ref
    +Ref/Only UDF    Data Type
      10/1     PR Preparative Data
      33/11    CR Chemical Reactions
      60/25    PP Physical Properties
       5/      DR Characterization Derivative
        /8     KW Short File Keywords
 2-bromo-thiophene
Preparative Data
   ...Preparation
      Starting Material: thiophene, cyanogen bromide
      Reagent: bromine
      Temp: 45 - 50 C
      By-product: 2.5-dibromo-thiophene (Ref. 6...
Refs.
   ... 6 , Steinkopf, Koehler, LACHDL, Liebigs Ann.Chem., 522
         (1936), 22
```

4E. REACTIONS DATA

The reactions data for 2-bromothiophene (33 reactions described and 11 others cited)
can be listed with a user-defined format of CR. This will list all of the reactions entries,
along with the relevant literature references:

```
?t 1/cr

 1/CR/1
  Data Present:
   Data  Ref
   +Ref/Only UDF    Data Type
     33/11    CR Chemical Reactions
Chemical Reactions
   Chemical Reaction:
      Partner: concentrated sulfuric acid
      Reaction Product: 2.5-dibromo-thiophene, bromo derivative of
      alpha, alpha-dithienyl, bromothiophenesulfonic acid (Ref. 2
      handbook)
   Chemical Reaction:
      Partner: fuming sulfuric acid
      Reaction Product: bromothiophenesulfonic acid, monobromo-
      dithienyl, dibromo-dithienyl (Ref. 2 handbook)
      .
      .
      .
      .
```

```
Chemical Reaction:
   Aim of the Study: Kinetic (Ref. 99)
Chemical Reaction:  (Ref. 100)
Chemical Reaction:  (Ref. 101)
Refs.
   2, Toehl, Schultz, CHBEAM, Chem.Ber., 27 (1894), 2835
   7, Steinkopf, LACHDL, Liebigs Ann.Chem., 430 (1923), 99
      .
      .
      .
      .
   100, Braun, Seelig, CHBEAM, Chem.Ber., 97(1964)3098
   101, Caspari, et al, RAACAP, Radiochim.Acta, 8(1967)102,104
```

The preparative data and the reactions data, along with the relevant sets of references, are listed in format 3.

4F. CHEMICAL DERIVATIVES

Chemical derivatives of the title compound are all identified in the DR field, which can be used to define a display format:

```
?t 1/dr

 1/DR/1
  Data Present:
   Data  Ref
   +Ref/Only UDF    Data Type
     5/       DR Characterization Derivative
Characterization Derivatives
    as 5-bromo-(2)thienylmercury(1+) chloride ( mp: 225 degree
    Celsius) (Ref. 102 handbook)
    as 5-bromo-(2)thienylmercury-acetate (mp: 134-135 degree
    Celsius) (Ref. 103 handbook)
    as 2-bromo-5-nitro-thiophene (mp: 48-49 degree Celsius): (Ref.
    104 handbook), (Ref. 105 handbook)
    as 5-bromo-thiophene-2-sulfonic acid amide (mp: 144 degree
    Celsius): (Ref. 106 handbook)
    as 5-bromo-thiophene-2-sulfonic acid anilide (mp: 94-95 degree
    Celsius): (Ref. 107 handbook)
   Refs.
     102, Hurd, Anderson, JACSAT, J.Amer.Chem.Soc., 75(1953)
     3517,3518
```

```
103, Nishimura, et al, Bl.Univ.Osaka Prefect., (A)6(1958)
127,128
104, Hurd, Anderson, JACSAT, J.Amer.Chem.Soc., 75(1953)
3517,3519
105, Steinkopf, et al, LACHDL, Liebigs Ann.Chem., 512(1934)
136,161
106, Steinkopf, et al, LACHDL, Liebigs Ann.Chem., 512(1934)
136,148
107, Terentew, Kadatskii, ZOKHA4, Zh.Obshch.Khim., 21(1951)
1524,1526;engl.Ausg.S.1667
```

4G. STRUCTURAL AND ENERGY PARAMETERS

Structural and energy properties, which include dipole moment and other measures of molecular energy, are all displayed in response to the SE format statement:

```
?t 1/se

 1/SE/1
Skeletal Characteristics
   Interatomic distances and angles (Ref. 22 handbook)
   Interatomic distances and angles (Ref. 23), (Ref. 24)
   Refs.
       22, Harshbarger, Bauer, ACCRA9, Acta Crystallogr.,(B)26(1970)
       1010,1014
       23, Karl, Bauer, ACBCAR, Acta Crystallogr.Sect.B, 28(1972)
       2619
       24, Harshbarger, Bauer, ACBCAR, Acta Crystallogr.Sect.B, 26
       (1970) 1010

Electrical Moment
   Dipole Moment: 1.39 Debye; Method: Dielectric cnst.; Solvent:
   benzene; (Ref. 25 handbook)
   Dipole Moment: 1.37 Debye; Method: Dielectric cnst.; Solvent:
   benzene; (Ref. 26 handbook)
   Dipole Moment: 1.36 Debye; Method: Dielectric cnst.; Solvent:
   benzene; (Ref. 27 handbook)
   Refs.
       25, Rogers, Campbell, JACSAT, J.Amer.Chem.Soc., 77(1955)4527
       26, Keswani, Freiser, JACSAT, J.Amer.Chem.Soc., 71(1949)218
       27, Nasarowa, Syrkin, ZOKHA4, Zh.Obshch.Khim., 23(1953)478;
       engl.Ausg.S.493
```

```
Electrical Polarizability
   Optical Anisotropy
      Optical Anisotropy:  (Ref. 28)
      Refs.
         28, Canselier, Clement, JCPBAN, J.Chim.Phys.Phys.
         Chim.Biol., 75(1978)880,882


Coupling Phenomena
   Coupling Constants
      Spin-Spin Coupling constants (Ref. 29), (Ref. 30), (Ref. 31)
      Refs.
         29, Takahashi, et al, JPCHAX, J.Phys.Chem., 74(1970)2765
         30, Read, et al, SPACA5, Spectrochim.Acta, 21(1965)85,89
         31, Hoffman, Gronowitz, ARKEAD, Ark.Kemi, 16(1961)515,
         519,520,523,531,532


Molecular Deformation & Potential
   Molecular Deformation
      Fundamental vibrations (Ref. 32 handbook), (Ref. 33
      handbook), (Ref. 34 handbook)
      Fundamental vibrations (Ref. 35), (Ref. 36), (Ref. 37),
      (Ref. 38)
      Refs.
         32, Garach, Lecomte, BSCFAS, Bull.Soc.Chim.Fr., 1946 423
         33, Hidalgo, JPRAAJ, J.Phys.Radium, (8)16(1955)366,368
         34, Katritzky, Boulton, JCSOA9, J.Chem.Soc., 1959 3500
         35, Horak, et al, SPACA5, Spectrochim.Acta, 22(1966)
         1355,1361
         36, Davidovics, et al, SAMCAS, Spectrochim.Acta Part A,
         23(1967)2669,2674
         37, Rabalais, et al, IJMIBY, Int.J.Mass Spectrom.Ion
         Phys., 9(1972)185
         38, Davidovics, et al, SAMCAS, Spectrochim.Acta Part A,
         23(1967)2669,2678


Molecular Energy Parameters
   Dissociation Energy:  Bond Type: C4H3S-Br; (Ref. 39 handbook)
   Refs.
      39, Szwarc, Williams, PRSLAZ, Proc.R.Soc.London, (A)219
      (1953)353,364
   Ionization Potential:  (Ref. 40)
   Refs.
      40, Fringuelli, et al, JCPKBH, J.Chem.Soc.Perkin Trans.2,
      (1976)276,277,278
```

4H. PHYSICAL STATE

A number of important properties are grouped together under the general heading of "physical state". These include the density, boiling point, and melting point and all crystal properties. All state of aggregation properties are displayed if the PS format is used:

```
?t 1/ps

 1/PS/1
Liquids
   Boiling Point: 153 - 154 C; Pressure: 760 Torr; (Ref. 10
   handbook)
   Boiling Point: 150.6 - 150.7 C; Pressure: 734.7 Torr; (Ref. 26
   handbook)
   Boiling Point: 70 C; Pressure: 50 Torr; (Ref. 41 handbook)
   Boiling Point: 63 C; Pressure: 38 Torr; (Ref. 25 handbook)
   Boiling Point: 43 - 44 C; Pressure: 15 Torr; (Ref. 8 handbook)
   Boiling Point: 38.8 - 39.5 C; Pressure: 13.5 Torr; (Ref. 27
   handbook)
   Boiling Point: 149 - 151 C; Pressure: 760 Torr; (Comment:
   Fluessigkeit.) (Ref. 1 handbook)
   Boiling Point: 151 - 151.5 C; Pressure: 760 Torr; (Ref. 7
   handbook)
   Boiling Point: 42 - 46 C; Pressure: 13 Torr; (Ref. 4 handbook)
   Boiling Point: 152 C; Pressure: 760 Torr; (Ref. 42)
   Boiling Point: 64 C; Pressure: 16 Torr; (Ref. 35)
   Boiling Point: 150 - 151 C; Pressure: 760 Torr; (Ref. 12)
   Boiling Point: 52 C; Pressure: 23 Torr; (Ref. 43)
   Boiling Point: 151 C; Pressure: 760 Torr; (Ref. 13)
   Boiling Point: 149 - 152 C; Pressure: 760 Torr; (Ref. 14)
   Boiling Point: 31 - 32 C; Pressure: 3 Torr; (Ref. 44)
   Boiling Point: 31 - 32 C; Pressure: 3 Torr; (Ref. 45)
   Boiling Point: 150 C; Pressure: 760 Torr; (Ref. 46)
   Boiling Point: 149 - 150 C; Pressure: 760 Torr; (Ref. 15)
   Boiling Point: 150 - 151 C; Pressure: 760 Torr; (Ref. 47)
   Boiling Point: 52 - 54 C; Pressure: 20 Torr; (Ref. 19)
   Boiling Point: 36 C; Pressure: 9 Torr; (Ref. 20)
   Boiling Point: 152 - 154 C; Pressure: 760 Torr; (Ref. 21)
   Boiling Point: 150 C; Pressure: 760 Torr; (Ref. 48)
   Refs.
      1, V Meyer, CHBEAM, Chem.Ber., 16 (1883), 1469, 1472
      4, Krause, Renwanz, CHBEAM, Chem.Ber., 62 (1929), 1710,
   CHBEAM, Chem.Ber., 65 (1932), 778
      7, Steinkopf, LACHDL, Liebigs Ann.Chem., 430 (1923), 99
      8, Terentew, et al, ZOKHA4, Zh.Obshch.Khim.,
```

```
24(1954)1265,1269;engl.Ausg.S.1251,1254
10, Blicke, Burckhalter, JACSAT, J.Amer.Chem.Soc.,
64(1942)477,478
12, Deganti, et al, BSFCAY, Boll.Sci.Fac.Chim.Ind.Bologna,
19(1961)76,77,81
13, Nemec, et al, CCCCAK, Collect.Czech.Chem.Commun.,
37(1972)3122
14, Pearson, et al, SYNTBF, Synthesis, (1976)621
15, Jones, Moodie, JSOOAX, J.Chem.Soc.C, (1960)2021
19, Trompen, Huisman, RTCPA3, Recl.Trav.Chim.Pays-Bas,
85(1966)175,179
20, Olsson, Axen, ARKEAD, Ark.Kemi, 22(1964)237,240
21, van der Plas, Persoons, RTCPA3,
Recl.Trav.Chim.Pays-Bas,
83(1964)701,704
25, Rogers, Campbell, JACSAT, J.Amer.Chem.Soc.,
77(1955)4527
26, Keswani, Freiser, JACSAT, J.Amer.Chem.Soc., 71(1949)218
27, Nasarowa, Syrkin, ZOKHA4, Zh.Obshch.Khim.,
23(1953)478;engl.Ausg.S.493
35, Horak, et al, SPACA5, Spectrochim.Acta,
22(1966)1355,1361
41, Motoyama, et al, NPKZAZ, Nippon Kagaku Zasshi,
78(1957)950,962, Zentralblatt: 1958 13204
42, Jeffery, et al, JCSOA9, J.Chem.Soc., (1961)570
43, Felloni, Pulidori, ANCRAI, Ann.Chim.(Rome),
51(1961)1027,1030
44, Lien, Kumler, JPMSAE, J.Pharm.Sci.,
59(1970)1685,1686,1688
45, Lien, Kumler, JPMSAE, J.Pharm.Sci.,
59(1970)1685,1686,1688
46, Clementi, Marino, JCPKBH, J.Chem.Soc.Perkin Trans.2,
(1972)71,72
47, Tundo, BSFCAY, Boll.Sci.Fac.Chim.Ind.Bologna,
18(1960)102,103
48, Hoffman, Gronowitz, ARKEAD, Ark.Kemi, 16(1961)515,534
```

V. Searching and Displaying Structures in Beilstein Online

Introduction and Purpose
The Beilstein Database
DIALOG Online Structure Search
Creating a ROSDAL String Manually
Creating a ROSDAL String Using MOLKICK
Graphics Output

5A. INTRODUCTION AND PURPOSE

This chapter describes what is in the Beilstein Structure Database and how it can be searched. Structure searching in the Beilstein Database is greatly facilitated by MOLKICK, a PC-resident structure drawing program, although it is possible to conduct structure searches without using MOLKICK. This section will not repeat what has been published in the MOLKICK manual, and for details concerning that program, users are urged to consult the manual.

Nomenclature searching is convenient and is widely used for searching for chemicals. With a chemical name, one can quickly find a specific compound if one has the name which is used in the database. In spite of this, nomenclature searching does have its limitations. The defining property of every chemical is its structure. No two chemical compounds have exactly the same total structure, although they may differ only in stereochemistry, or in the presence or absence of an isotope or a charge. Structure searching has become the predominant search mode for chemicals. This chapter will describe the full and partial structure searching capabilities available on DIALOG's Beilstein Database.

5B. THE BEILSTEIN DATABASE

The Beilstein Structure File, as implemented on DIALOG, contains for most compounds the IUPAC name (CN), the Beilstein Registry Number (BN), and the chemical structure connection table in searchable form. All compounds in the Beilstein Structure File are carbon compounds and have factual data records associated with them. The Beilstein classification system does not involve, nor does it need, the creation of imaginary compounds in order that the indexing system function correctly. Inorganic compounds, which are not in the *Beilstein Handbook* or the Beilstein Structural and Factual Databases, are to be found in the *Gmelin Handbook*, which is expected to be available online in the early 1990s. Chemical Abstracts Service (CAS) Registry Numbers (RN) are available for many of the compounds in the Beilstein Database and can be used for cross-file searching between the Beilstein Database (file 390) and the CAS bibliographic files (308–312).

This chapter deals with only the contents of the Beilstein Registry connection table file, which will be referred to as the Beilstein Structure File. This file is being implemented in parts; when it is complete, it is expected to contain over five million chemical compounds, together with the associated factual data, all of which will be available online. It will continue to be useful to have access to the printed *Handbook*, which currently comprises over 350 volumes. This is because some records in the online database are extraordinarily voluminous, and also because data from the last 10–20 years, while in the *Handbook*, may not yet be online.

It should be remembered that the only compounds in the Beilstein Structural and Factual Databases are organic compounds for which there is an accurate description, which requires evidence as to purity, known structure, and accurately reported chemical and physical properties. It is also important to remember that while there are, at the beginning of 1990, some three million compounds in the factual database, very few, if any, have data in all possible fields. As noted in the previous chapter, the "data present" (DP=) search qualifier allows one to see what data are actually in the database for the compound or compounds of interest.

The Beilstein Database will be updated rather infrequently. The update cycle is currently about six months.

The Beilstein Registry structure file, as implemented on DIALOG, consists of a variety of fields which can be used for search and display on a structural or substructural basis. These fields are:

1 *Beilstein Registry Number (BN).*   This is analogous to the CAS Registry Number. It is a unique number used to identify a chemical compound and the record associated with it. The lowest BN is 1001, and the BNs in the current database range up to about 3000000. There are no hyphens in Beilstein Registry Numbers. The order of the entire database is by descending Beilstein Registry Number. Thus in any display, the most recently reported compounds (highest BNs) will appear first; the oldest, last.

2 *ROSDAL string.*   This is a mathematical representation of a chemical structure, written in a form which can be used by computer programs to conduct structure or substructure searches or to display the structure on a graphics terminal or a printer. Within the ROSDAL string, every atom is identified in terms of its elemental nature and the types of bonds by which it is connected to other atoms in the molecule.

3 *Atom labels.*   These are the pieces of information in the ROSDAL string which allow one to specify details about each atom. These labels include:

- *Element type.*   Standard symbols are used.

- *Charge.*   Default is zero; otherwise specified, with sign.

- *Free sites.*   Default is zero. If attached atoms, beyond those shown in the diagram, are required, the number of free sites at an atom must be specified.

- *Other Specifications.*   These are less frequently used attributes, such as abnormal valency or mass.

- *Lock symbol.*   This allows one to toggle between the single specified value to the atom (e.g. C) and a generic value (e.g. C, O, or N).

5C. DIALOG ONLINE STRUCTURE SEARCH

The S4 search software used by DIALOG is designed to search chemical databases of structures which are represented as ROSDAL strings. Searches can be conducted for exact structures, for substructures, and for generic substructures, in which a good deal of indeterminate structural information may be present.

In order to conduct any of these searches, the DIALOG system must be presented with the appropriate ROSDAL string, which is essentially the argument that follows the search qualifier, QS=. This ROSDAL string may be generated manually, as is described later in the next section of this chapter, or it may be generated with MOLKICK, a PC-based program which can be resident in PC memory at the same time that DIALOG is being accessed.

Completion of a QS= search, no matter how it was begun, leads to the creation of a *set* of answers—records of compounds which were retrieved. Such sets are just like any sets obtained from searches in DIALOG. They can be displayed, typed, printed, or combined (AND, OR, NOT) with independently derived sets.

5D. CREATING A ROSDAL STRING MANUALLY

5D1. ROSDAL Strings

ROSDAL (Representation of Organic Structure Descriptions Arranged Linearly) is a method of describing a chemical structure query by means of a string of alphanumeric characters. This description of organic structures is used in the Beilstein Database to begin structure or substructure searches.

There are two methods of passing structural information into and out of the Beilstein Database on DIALOG. The first of these is via ROSDAL strings, entered at the keyboard, and the second is by means of MOLKICK, which is described in the next chapter and whose use is described in the next section of this chapter. MOLKICK is a PC-resident program which allows one to draw chemical structures using a mouse. Once a structure is drawn, MOLKICK will convert it to the corresponding ROSDAL string, which is then passed to the search system for a structure or substructure search. If one uses MOLKICK, it is not necessary to know how ROSDAL strings are assembled, but such knowledge is needed if ROSDAL strings are going to be entered from the keyboard. The next section in this chapter explains how ROSDAL strings can be generated manually, and the subsequent section describes the use of MOLKICK.

5D2. Structure and Substructure Searching with ROSDAL Strings

ROSDAL strings are very versatile and powerful; they support the linear, alphanumeric description of all types of organic molecular structures, including full structures, partial structures, and generic structures.

A *full structure* is one in which all unfilled valencies are occupied by hydrogen, i.e., the molecule is represented in the conventional way. A *partial structure*, or substructure, is a structure in which one or more bonds may be incompletely defined, unoccupied valencies are allowed, and atoms of variable type are also allowed. *Generic structures* are those which may contain generic groups, such as alkyl, which can assume a large number of identities. In the three structures

1 is a full structure. It describes a single compound, and no variability of any sort is possible. Structure 2 is a substructure, because attachments of any sort are allowed at the atoms marked with the * (these are termed *free sites*). Structure 3 is a generic structure, because not only are there several free sites, but the atom X may be any of a number of (user-defined) types.

In principle, a full structure search in the Beilstein Database will only retrieve a single compound. In practice however, because the ROSDAL strings do not carry data pertaining to stereochemistry, unusual isotopes, or charges, such modifications of the single structure may be retrieved simultaneously.

A substructure search will retrieve all compounds in which the query structure can be imbedded. The conditions of imbedment can be controlled by the user, who can, for example, alter the number and position of free sites. In general, the simpler the substructure, the larger the number of retrievals, and with extremely simple and common substructures, such as an unsubstituted phenyl ring, the substructure search will fail because it exceeds a system limit (too much time or too many hits). Generic structure searches can also be prohibitively open ended, and the query structures used in such searches must be developed with some care in order that a useful and manageable set of retrievals will result from their use.

5D3. Manual Generation of ROSDAL Strings

The generation of the ROSDAL string corresponding to *p*-chlorotoluene will be described in this section. Whether this is a full structure, a substructure, or a generic structure is determined by the content of the ROSDAL string that is generated, as will be seen.

Development of a ROSDAL string involves five steps, which are described below:

1   Draw the structural formula in the usual way, and then number the atoms. The

p-chlorotoluene example is shown below:

You should not assign the same number to more than one atom, but otherwise, you may use any numbers you wish. Hydrogen atoms should be ignored when numbering unless they are bridging, charged, or isotopically labeled.

2   Write a string that describes the connectivity of the structure, i.e. what is connected to what. Pay no attention at this stage to the type of bond; presence of a bond is all that is required here. Each atom need only be specified once, but multiple specification is permitted. If the structure was numbered as shown above, this will lead to the following connectivity:

$$1-2-3-4-5-6-1,1-7,4-8$$

This can be legally expressed in other ways, e.g.

$$1-2-3-4-5-6-1-7,4-8$$
$$1--6-1-7,4-8$$

Shortcuts exist to describe a string of atoms. Thus six carbons in a chain with only single bonds (*n*-hexane) gives a ROSDAL of $1-2-3-4-5-6$, which can be abbreviated $1--6$. A benzene ring can be written as $1-2=3-4=5-6=1$, abbreviated as $1-=6=1$.

3   Define the atom types. This is done by appending the element symbol to the atom number. Thus atom 8, a chlorine, must be rewritten as 8Cl:

$$1-2-3-4-5-6-1,1-7,4-8Cl$$

If an atom is not defined, it is assumed to be carbon. Thus every atom except atom 8 in the above string will be a carbon.

4   Define the bond types. In the string shown above, all bonds are written as single bonds. To reflect the *p*-chlorotoluene structure accurately, alternating single and double bonds must be placed around the ring:

$$1=2-3=4-5=6-1,1-7,4-8Cl$$

5   Terminate the ROSDAL string with a period (.):

$$1=2-3=4-5=6-1,1-7,4-8Cl.$$

This string is now complete, and if it is passed to DIALOG as a QS (query structure), seven hits result:

```
?qs

QS/QC Query Structure Version 1.01
Enter ROSDAL connection table
Line No.  1 (enter '.a' to end or 'ABORT' to quit)
?1=2-3=4-5=6-1,1-7,4-8Cl.
Line No.  2 (enter '.a' to end or 'ABORT' to quit)
?aa

DASD 2FF  DETACHED
Processing - Atom-by-atom match started
        S1        7  QS  ID Rosdal: 1=2-3=4-5=6-1,1-7,4-8CL
```

These are arranged in descending order of Beilstein Number, and accordingly, the last one (the lowest Beilstein Number) is *p*-chlorotoluene:

```
?t/bi/7

 1/BI/7
1903635
  Lawson No: 4109
  Beilstein Cit: 5-05
  Molecular Formula: C7H7Cl
  Molecular Weight: 126.59
  Synonym: p-Chlortoluol
  No. of Ref: 349
  Data Present:
   Data  Ref
   +Ref/Only UDF    Data Type
      /7    PR Preparative Data
    92/91   CR Chemical Reactions
    36/103 PP Physical Properties
     1/     PB Physiological Data
     1/     DR Characterization Derivative
      /12   KW Short File Keywords
```

The first five compounds are isotopically labeled *p*-chlorotoluene, and the sixth compound is the incomplete molecule *p*-chlorobenzyl.

TABLE XXV

SYMBOL	VALUE
Q	Any atom, excluding C or H
QH	Any atom, excluding C
M	Any metal
MH	Any metal or H
X	Any halogen (F, Cl, Br, I)
XH	Any halogen or H

5D4. Structure Variations in ROSDAL Strings

The signs that a ROSDAL string corresponds to a substructure or a generic structure are placed within the atom attributes, which is where the atom type was defined. An atom attribute can be used to define the atom type, a free site, or an attachment point, as well as charge, mass, valency, or radical nature.

5D4A. Atom Type

For the atom type, any elemental symbol may be used. With no symbol, the atom is assumed to be carbon. In addition to element symbols, the generic atoms **A** and **AH** (any atom, excluding H, and any atom, including H) can be used. The symbol **AH** may be further qualified by a numeric range which indicates, inclusively, the atomic numbers of the allowed elements. Thus **AH<7,8>** means that only N and O are allowed.

A variety of *generic elements* are available, as detailed in Table XXV. The term metal means any element except Ar, As, At, B, Br, C, Cl, F, H, D, T, He, I, Kr, N, Ne, O, P, Rn, S, Se, Si, Te, and Xe.

Any element symbol may be negated (i.e. its *absence* required) if the corresponding symbol is preceded by a tilde (~). Thus if **X** signifies that a halogen must be present, ~**X** will signify that a halogen must not be present, but any other element may be.

A variety of *generic element symbols*, shown in Table XXVI, are available and can be used to assign variable values to any atom in the query structure. This table uses the following definitions:

- Acyclic: A group containing no rings.

- Alkenyl: A group consisting of only C and H atoms and containing one or more double bonds; rings and triple bonds are not allowed.

- Alkoxy: An —O— alkyl group.

- Alkyl: A group consisting of only C and H atoms in which all bonds are single; rings and multiple bonds are not allowed.

- Alkynyl: A group consisting of only C and H atoms and containing one or more triple bonds; double bonds are also allowed, but rings are not allowed.

- Aryl: A group consisting of at least one optionally substituted aromatic ring system which is connected directly to the parent structure.

- Carbacyclic: An acyclic group consisting only of C and H.

- Carbocyclic: A group consisting of only C and H atoms and containing one or more rings.

- Cyclic: A group containing one or more rings.

- Cycloalkenyl: A group consisting only of C and H atoms and containing one or more rings connected directly to the parent structure; at least one ring must contain at least one double bond.

- Cycloalkyl: A group consisting only of C and H atoms and containing one or more rings connected directly to the parent structure; multiple bonds within rings are not allowed.

- Heteracyclic: An acyclic group containing one or more heteroatoms.

- Heteroaryl: A group containing at least one (optionally substituted) fused aromatic ring system which is connected directly to the parent structure and which contains at least one heteroatom.

- Heteroatom: Any atom other than H, D, T, or C.

- Heterocyclic: A group containing one or more rings and in which at least one ring contains one or more heteroatoms.

Residue group symbols (G0, G1, G2,..., G99) may be used in generic query structures to identify a group of alternative substituents. Residue group symbols may be defined using any combination of structural formulas and generic element symbols.

If a residue group symbol G0 is to be used in a query structure, the string is first written containing the G0 at the correct place. Thus 1-G0-2 signifies a three atom chain whose middle atom is to be generically defined. The definition of G0 is then appended to this string, so the whole ROSDAL string might be

$$1-G0-2,G0=(10;1N\&)$$

which indicates that a dimethyl ether or a di- or trisubstituted dimethylamine is expected.

Residue groups (G-groups) may not themselves contain free sites, and they may not be defined recursively (i.e., they may not contain themselves). Nesting of G-groups is allowed, however. So while the definition of G1 may not itself contain G1, it may contain G2.

TABLE XXVI

SYMBOL	VALUE
G	Any group, excluding H
GH	Any group or H
G*	Any group, excluding H, with free site(s)
GH*	Any group or H, with free site(s)
ACY	Acyclic group
ACH	Acyclic group or H
ABC	Carbacyclic group
ABH	Carbacyclic group or H
AHC	Heteracyclic group
AHH	Heteracyclic group or H
ALK	Alkyl group
ALH	Alkyl group or H
AEL	Alkenyl group
AEH	Alkenyl group or H
AYL	Alkynyl group
AYH	Alkynyl group or H
AOX	Alkoxy group
AOH	Alkoxy group or H
CYC	Cyclic group
CYH	Cyclic group or H
CBC	Carbocyclic group
CBH	Carbocyclic group or H
CHC	Heterocyclic group
CHH	Heterocyclic group or H
CAL	Cycloalkyl group
CAH	Cycloalkyl group or H
CEL	Cycloalkenyl group
CEH	Cycloalkenyl group or H
ARY	Aryl group
ARH	Aryl group or H
HAR	Heteroaryl group
HAH	Heteroaryl group or H

A G-group may be connected to a maximum of seven other atoms. Each G-group msut be defined in turn within parentheses at the end of the ROSDAL string and preceded by a semicolon (;), which marks the end of the previous structural group or G-group:

$$\ldots;G0=(\ldots);G1=(\ldots);G2=(\ldots).$$

If it is wished to specify substitution at one or more of several positions in a query structure, the G-group attached to each of the possible positions and the desired frequency of occurrence must be specified. The frequency is a maximum value; a minimum value of 1 is assumed. Thus

$$\ldots;G0=3*(\ldots).$$

specifies mono-, di-, and trisubstitution of the structure by G0.

5D4B. Free Sites

A complete structure (no generic atoms, G-groups, or free sites) will give only a full structure search, as was illustrated in section 5D3. Incorporation into such a structure of *free sites*, or points of allowable attachment of other atoms, will convert it to a substructure, and use of this query structure in a search will lead to a substructure search. Thus the ROSDAL string that was developed earlier,

$$1=2-3=4-5=6-1,1-7,4-8Cl$$

will lead to a full structure search for *p*-chlorotoluene:

```
?qs

QS/QC Query Structure Version 1.01
Enter ROSDAL connection table
Line No.  1 (enter '.a' to end or 'ABORT' to quit)
? 1=2-3=4-5=6-1,1-7,4-8Cl.
Line No.  2 (enter '.a' to end or 'ABORT' to quit)
? aa

DASD 2FF   DETACHED
Processing - Atom-by-atom match started
      S2       7  QS  ID Rosdal: 1=2-3=4-5=6-1,1-7,4-8CL
```

If a free site is specified within the string, as in

$$1=2*1-3=4-5=6-1,1-7,4-8Cl$$

the ensuing search will retrieve all *p*-chlorotoluenes containing (optionally) a substituent of any sort at the ring carbon next to the methyl group:

```
?qs

QS/QC Query Structure Version 1.01
Enter ROSDAL connection table
Line No.  1 (enter '.a' to end or 'ABORT' to quit)
? 1=2*1-3=4-5=6-1,1-7,4-8Cl.
```

```
Line No.  2 (enter '.a' to end or 'ABORT' to quit)
? aa


DASD 2FF  DETACHED
Processing - Atom-by-atom match started
     S2      573  QS  ID Rosdal: 1=2*1-3=4-5=6-1,1-7,4-8CL
```

Note that many more hits (573 versus 7) result from the addition of this free site. Note also that the search leading to S2 is a *substructure search* and costs more than the full structure search.

The free site designation means only that one or more nonhydrogens may be attached at the atom in question. Nothing is implied about the nature of the attachment. The number of attachments may be controlled, and while *1 means that 0 or 1 attachments are permitted, *<1,2> means that 1 and 2 are the only allowed values; 0 and 3,4,... will not be permitted.

Free sites may also be used with the generic group symbols G and GH. The symbol G* or GH* means that the G-group is allowed to form a ring with any other atom or G-group that also has a free site.

Generic structures may use *attachment points*, which allow specification of how a substituent group is to be attached to its parent structure. If the substituent is attached via only one atom, then the attachment point symbol (&) should be specified as an attribute of the atom. Thus a *n*-propyl group could be written as 1&-2-3, but the isopropyl group would be 1-2&-3. The nature of the attachment bond is specified in the description of the parent structure. If the substituent consists of a single atom, it is not necessary to specify the attachment point.

A substituent may be attached to the parent structure via more than one atom (simultaneously or not); then an attachment point symbol must be written on each of the appropriate atoms of the substituent. Thus if G0 is defined as 1&-2N&, then (parent)-N-C and (parent)-C-N will both be allowed.

If it is wished to restrict how a substituent is attached to its parent, then each attachment point symbol in the substituent must be followed with the atom number of the node in the parent structure to which that atom should be attached. For example, if the parent structure is

$$-7-G0-9-$$

and G0 is (1C&7-2N&9), then only 7-C-N-9 will be implied. If G0 is (1C&9-2N&7), only 7-N-C-9 will be retrieved.

An atom attribute may also contain the *charge* on the atom, specified as (+n) or (-n). If no charge is specified, both charged and uncharged structures will be retrieved.

The *mass* of an atom may be specified as an attribute to the atom, e.g.

$$...8CL(W36)....$$

TABLE XXVII

SYMBOL	VALUE
-	Single
=	Double
#	Triple
?	Any bond
!	Cut bond

The default is unspecified, i.e., all masses will be retrieved. Standard atomic masses need not be specified unless they are to be the only mass allowed.

The *valency* of any atom may also be specified within the atom attribute, as in ...6N(V5).... A charged pentavalent nitrogen would be ...6N(+1,V5).... If there is no valency specification, all values of valency at that atom will be allowed.

The *radical* specification refers to the number of unpaired valence electrons on an atom. If there is no radical specification, then all radical forms will be retrieved.

These atomic attributes are currently not implemented in the substructure search software and if cited in a structure query, they will be ignored.

5D4C. Bond Types

The *bond values* that can be used in ROSDAL strings are given in Table XXVII. The symbol for a "cut bond" (!) may be used to delete a bond that was specified earlier in the ROSDAL string.

Additional *attributes to bonds* can be added to the ROSDAL string after the bond symbol and placed in parentheses, e.g. (R). If more than one such attribute is used, they should be separated from one another with commas, e.g. (R,=), which means that the bond in question should be double and in a ring.

The *bond type* supports specification of a ring or chain environment for the bond. A bond of type R must be in a ring, one of type C in a chain, and one of type RC may be in either a ring or a chain. These attributes are not currently implemented.

5D5. Examples of Manually Generated ROSDAL Strings

This section gives some examples of query structures, together with the corresponding ROSDAL strings.

1   *t*-Butyl chloride:

ROSDAL string:

$$1\text{-}2\text{-}3,2\text{-}4,2\text{-}5Cl. \quad \text{or} \quad 1\text{--}3,2\text{-}4,2\text{-}5Cl.$$

```
?qs

QS/QC Query Structure Version 1.01
Enter ROSDAL connection table
Line No.  1 (enter '.a' to end or 'ABORT' to quit)
?1-2-3,2-4,2-5Cl.
Line No.  2 (enter '.a' to end or 'ABORT' to quit)
?aa

DASD 2FF  DETACHED
Processing - Atom-by-atom match started
      S3      17  QS  ID Rosdal: 1-2-3,2-4,2-5CL
```

The alternative ROSDAL string gives the same result:

```
?qs

QS/QC Query Structure Version 1.01
Enter ROSDAL connection table
Line No.  1 (enter '.a' to end or 'ABORT' to quit)
?1--3,2-4,2-5cl.
Line No.  2 (enter '.a' to end or 'ABORT' to quit)
?aa

DASD 2FF  DETACHED
Processing - Atom-by-atom match started
      S4      17  QS  ID Rosdal: 1--3,2-4,2-5CL

?t /cn sy/1-17

 3/CNSY/1
  Synonym: Perdeutero-tert-butylchlorid

 3/CNSY/2
  Synonym: beta-Chlor-tert.-butyl-Radikal

 3/CNSY/3
  Synonym: (37Cl)-Perdeutero-tert-butylchlorid

 3/CNSY/4
  Synonym: Nonadeuterio-tert-butylchlorid
```

```
3/CNSY/5
 Synonym: (35Cl)-Hexadeutero-tert-butylchlorid

3/CNSY/6
 Synonym: t-Butylchlorid-d9

3/CNSY/7
 Synonym: 1,1,1,3,3,3-Hexadeutero-2-chlor-2-methylpropan

3/CNSY/8
 Synonym: tert.-Butyl chloride-2-13 C-37 Cl

3/CNSY/9
 Synonym: tert.-Butyl chloride-2-13 C-35 Cl

3/CNSY/10
 Synonym: tert.-Butyl chloride-1-13 C

3/CNSY/11
 Synonym: 2-Chlor-2-methyl-1,1,1-trideuterio-propan

3/CNSY/12
 Synonym: 2-Chlor-2-methyl-1,1-dideuterio-propan

3/CNSY/13
 Synonym: tert.-Butylchlorid

3/CNSY/14
 Synonym: tert.-Butylchlorid

3/CNSY/15
 Synonym: 13 C-tert.-Butyl chloride

3/CNSY/16
 Synonym: 2-Chlor-2-methyl-1-deuterio-propan

3/CNSY/17
2-chloro-2-methyl-propane
```

2   Monodeuterated *t*-butyl chloride:

ROSDAL string:

1-2-3-4(W2),2-5,2-6Cl. or 1--4(W2),2-6,2-6Cl.

```
?qs


QS/QC Query Structure Version 1.01
Enter ROSDAL connection table
Line No.  1 (enter '.a' to end or 'ABORT' to quit)
?1--4H(w2),2-5,2-6cl.
Line No.  2 (enter '.a' to end or 'ABORT' to quit)
?aa


DASD 2FF  DETACHED
Processing - Atom-by-atom match started
      S5       17  QS  ID Rosdal: 1--4H(W2),2-5,2-6CL


?c 3 and 5
               17  3
               17  5
      S6       17  3 AND 5
```

Note that this search gives the same number of retrievals as the search for t-butyl chloride on the previous page. This is because the search software currently ignores atom attributes such as mass.

3 Bromobenzene:

ROSDAL string:

1-2=3-4=5-6=1-7Br. or 1-=6-7Br.

```
?qs


QS/QC Query Structure Version 1.01
Enter ROSDAL connection table
Line No.  1 (enter '.a' to end or 'ABORT' to quit)
?1-2=3-4=5-6=1-7Br.
Line No.  2 (enter '.a' to end or 'ABORT' to quit)
?aa


DASD 2FF  DETACHED
Processing - Atom-by-atom match started
      S7       24  QS  ID Rosdal: 1-2=3-4=5-6=1-7BR
```

4   3-Bromopyridine:

ROSDAL string:

1=2-3=4N-5=6-1-7Br   or   1=-6

```
?qs

QS / QC Query Structure Version 1.01
Enter ROSDAL connection table
Line No.  1 (enter '.a' to end or 'ABORT' to quit)
?1 = 2 - 3N = 4 - 5 = 6 - 1 - 7Br.
Line No.  2 (enter '.a' to end or 'ABORT' to quit)
?aa

DASD 2FF  DETACHED
Processing - Atom-by-atom match started
      S8        3  QS  ID Rosdal: 1 = 2 - 3N = 4 - 5 = 6 - 1 - 7BR

? t /cn mf sy / 1 - 3

 8 / CNMFSY / 1
1363781
  Molecular Formula: C5H5BrN
  Synonym: 3 - Brom - pyridinium

 8 / CNMFSY / 2
387746
  Molecular Formula: C5BrD4N
  Synonym: 3 - Brompyridin - d4

 8 / CNMFSY / 3
105880
3 - bromo - pyridine
  Molecular Formula: C5H4BrN
```

5   Substituted nitrobenzenes:

```
?qs

QS/QC Query Structure Version 1.01
Enter ROSDAL connection table
Line No.  1 (enter '.@' to end or 'ABORT' to quit)
?1n-=7-2,1=80,1=90,3*,4*,5*,6*,7*.
Line No.  2 (enter '.@' to end or 'ABORT' to quit)
?@@


DASD 2FF  DETACHED
Processing - Atom-by-atom match started
      S9  166767  QS  ID Rosdal: 1N-=7-2,1=80,1=90,3*,4*,5*,
                                 6*,7*.


?t /gs bi/50


 9/GSBI/50
3123905
   Lawson No: 16437, 13415, 289
   Beilstein Cit: 5-15
   Molecular Formula: C18H20N4O7
   Molecular Weight: 404.38
   Synonym: 8 a-Hydroxy-2a-methoxycarbonyl-adamantanon-
            (4)-2.4-dinitrophenylhydrazon
   No. of Ref: 1
Graphic Structure:
'3123905'
```

```
Data Present:
 Data  Ref
 +Ref/Only UDF    Data Type
   2/       PP Physical Properties
```

6 2,4-Dinitrophenylhydrazones:

ROSDAL String:

1*2=2N-3N-4-5=6-7=8-9=4,7-10N=110,10N=120,5-13N=140,13N=150.

```
?qs

QS/QC Query Structure Version 1.01
Enter ROSDAL connection table
Line No.  1 (enter '.@' to end or 'ABORT' to quit)
?1*2=2n-3n-4-5=6-7=8-9=4,7-10n=11o,10n=12o,5-13n=14o,13n=15o.
Line No.  2 (enter '.@' to end or 'ABORT' to quit)
?@@


DASD 2FF  DETACHED
Processing - Atom-by-atom match started
Processing - atom by atom 50% complete 7647 hits 9093
            estimated total
Processing - atom by atom 75% complete 20009 hits 21501
            estimated total
    S10     20359  QS  ID Rosdal: 1*2=2N-3N-4-5=6-7=8-9=4,
                        7-10N=110...


?t /bi gs/5-8

 10/BIGS/5
3124376
  Lawson No: 16437, 7329
  Beilstein Cit: 5-15
  Molecular Formula: C34H50N4O4
  Molecular Weight: 578.79
```

Synonym: 6 beta-Methyl-Delta4-cholesten-3-on-
          2.4-dinitro-phenylhydrazon
  No. of Ref: 1
Graphic Structure:
'3124376'

  Data Present:
   Data Ref
   +Ref/Only UDF   Data Type
      2/   PP Physical Properties

 10/BIGS/6
3124324
  Lawson No: 16437, 7748
  Beilstein Cit: 5-15
  Molecular Formula: C19H18N8O8
  Molecular Weight: 486.4
  Synonym: trans-1.2-Diacetyl-cyclopropan-bis-
        (2.4-dinitrophenylhydrazon)
  No. of Ref: 1
Graphic Structure:
'3124324'

```
  Data Present:
   Data  Ref
   +Ref/Only UDF    Data Type
      1/    PP Physical Properties
       /1   KW Short File Keywords

 10/BIGS/7
3124323
  Lawson No: 16437, 7748
  Beilstein Cit: 5-15
  Molecular Formula: C19H18N808
  Molecular Weight: 486.4
  Synonym: cis-1.2-Diacetyl-cyclopropan-bis-
           (2.4-dinitrophenylhydrazon)
  No. of Ref: 1
Graphic Structure:
'3124323'
```

```
  Data Present:
   Data  Ref
   +Ref/Only UDF    Data Type
      1/    PP Physical Properties
       /1   KW Short File Keywords

 10/BIGS/8
3124264
  Lawson No: 16437, 13813, 8502
  Beilstein Cit: 5-15
  Molecular Formula: C25H32N407S
  Molecular Weight: 532.61
  Synonym: 2t-Methyl-2c-p-toluolsulfonyloxymethyl-
           4r-t-butyl-cyclohexanon-2.4-dinitro-
           phenylhydrazon
```

```
   No. of Ref: 1
Graphic Structure:
'3124264'
```

```
Data Present:
 Data   Ref
 +Ref/Only UDF     Data Type
    1/        PP Physical Properties
```

5E. CREATING A ROSDAL STRING USING MOLKICK

5E1. Introduction to MOLKICK

MOLKICK is a program which runs on PCs and which supports the graphical generation of chemical structures and their uploading as query structures to online search
systems such as DIALOG, STN, and QUESTEL.

An important feature of MOLKICK is that it is memory resident and can be
executed at the same time you are using the PC to communicate with the online
system. In this way, query structures may be rapidly modifed and resubmitted for
searches, and the feedback that is possible allows for very efficient structure and
substructure searching.

A very detailed description of MOLKICK will be found in the *MOLKICK User's
Manual* (1). This section does not attempt to work at that level; rather it is a brief
description of the program, in the context of the Beilstein Database on DIALOG.

5E2. Installation of MOLKICK

MOLKICK can be run on any IBM or IBM-compatible PC which is equipped with a
graphics driver and a mouse. A separate directory should be established for the
MOLKICK programs and files. The program requires approximately 500 K of memory;
this varies with the query structure that is being generated. With a standard 640 K PC,
however, there is sufficient memory for the simultaneous installation of MOLKICK and
a communications package such as DIALINK, PCPLOT, or CROSSTALK.

When a query structure has been completed in MOLKICK, it can be converted (function key F7) to a ROSDAL string, which is stored locally. At this stage, one can exit from MOLKICK, connect to DIALOG, and then, with ALT-R, upload the stored ROSDAL string to the Beilstein Database Search System and begin the structure search.

Results from a search on DIALOG can be passed back to MOLKICK, which can display structures of compounds in the Beilstein Database. MOLKICK, however, cannot display both structures and text; a separate program, GEOFF (see section 5E4) is very useful for this purpose.

It is generally not necessary to deinstall MOLKICK during a DIALOG session, but the program can be deinstalled, and the memory released, at any time with a CTRL-END command.

5E3. Use of MOLKICK

MOLKICK is almost entirely mouse driven. The working screen has eight commands for structure drawing. Each of these, when picked by the mouse, produces a pulldown menu, which offers further choices. The function keys F1–F10 are all active and support MOLKICK commands for control of the program.

5E3A. Hot Key

When MOLKICK is installed in the PC memory, it can be turned on and off with ALT-SHFT, which is called the *hot key*. Both keys should be pressed simultaneously; then the MOLKICK screen will appear. If the MOLKICK screen is present, ALT-SHFT will cause it to disappear, and the PC will be left at the DOS level in the MOLKICK directory.

The hot key does not deinstall MOLKICK; this should be done when you have finished your session and is accomplished with the CTRL-END key combination. When this is done, the ALT-SHFT combination no longer works, and MOLKICK must be called again from its directory if it is to be reinstalled.

5E3B. Structure Generation

When the MOLKICK screen comes up, the program is in structure-drawing mode. Clicking the left button of the mouse deposits an atom on the screen where the cursor is. Movement of the mouse and another left button click will create a bond from the first atom, and then a second atom. To release an atom, i.e. move away from it without drawing a new bond, the right button of the mouse should be clicked. By default, all atoms will be C and bonds will be single. Both however can be changed easily.

5E3C. Modification of Query Structures

When the required skeleton is complete, atoms can be modified as desired. To change an atom's identity, move the cursor onto the atom and click the left button. A pulldown menu appears containing the current attributes of that atom (number, symbol, charge, free site, other specifications, and lock symbol). Any of these can be changed by moving the cursor to the appropriate place and then typing in the new value. To define an atom as nitrogen, for example, left click on that atom in the query structure, move to the "symbol" line of the pulldown menu, and overwrite the C that is there with an N.
   Modification of bonds requires a similar exercise. To change a single bond to a double bond, begin by typing 2 to signify that you wish a double bond. Then left click on one of the atoms, then the other. The existing (single) bond will begin to flash, and a third click on the left button will convert it to a double bond. Finally, move the cursor off the atom and right click to detach it from the structure.

5E3D. Free Sites

It is important to remember that in contrast to STN, MOLKICK establishes no free sites automatically. Unless free sites are designated by the user, a search with a structure developed by MOLKICK will be a *full structure search*. If you wish to do a substructure search, you must identify one or more free sites in the query structure. This can be done much as atoms are identified. A left click on the atom in question produces its pulldown menu, whose fourth line is the "free site" line. The default value for the free sites is zero. If you wish to allow up to *n* attachments in addition to those already at that atom, enter *n* as the free site value. This must be done at every atom where attachments are to be allowed.

5E3E. Generic Query Structures

Query structures generated by MOLKICK may have the same types of generic features as those described in section 5D4 in manually generated ROSDAL strings. Any atom in the query structure may be identified by its element symbol as that element. In addition, the full range of generic symbols in Table XXVI in section 5B4A may be used in place of the standard element symbol. In this way, it is possible to define a query structure which, at some point, has any heterocyclic ring, or any halogen atom attached to it.
   Markush structures, in which an attached group may have any of several different structures, can be defined by MOLKICK. This is done by placing an atom attachment at the atom in question and defining the new atom as G1 (or G2, G3, ...). This will cause the existing (mother) structure to be put into an inset on the screen. The working screen will be cleared, and you may now use MOLKICK to develop a definition for G1. This G1 may itself contain G-structures, and the process will be repeated until the definition is complete.

5E3F. Function Keys

The function keys F1–F10 (with or without SHFT) store MOLKICK commands, as described below.

- F1: HELP key. This is context sensitive. When it is pressed, it will provide you with assistance concerning the part of MOLKICK you are currently using. You can page through help messages with the cursor keys. Pressing F1 a second time removes the help message from the screen.

- F2: Detects and removes overlapped atoms. When building a query structure, atoms may sometimes overlap one another on the screen. This key will detect such conditions and highlight them on the screen. If there are no overlaps, a beep will be heard. To remove overlaps, you may move, delete, or merge atoms with the mouse as described in the *MOLKICK Manual*.

- F3: This key is a toggle switch which may be used when editing nested Markush structures. It turns on and off the display of the next higher structure.

- F4: This key is used when working on Markush structures. When editing a nested Markush structure, it returns you to the mother structure.

- F5: This key allows you to "undo" (nullify) the last command. It can be used to recover from mistakes.

- F6: KILL key. This key erases your current structure. It can be reversed with F5. If you are working with a Markush structure, F6 will erase the actual residue and all lower level structures. To erase an entire Markush structure, use F4, then F6.

- F7: This key stores the structure on the screen as a ROSDAL string in a file named SUB.UPL. Each time you press F7, the old SUB.UPL is renamed SUB.BAK and a new SUB.UPL is created. Thus the two most recent versions are always available for editing or uploading.

- F8: Unused.

- F9: This key creates a CAS-ONLINE search string from the structure on the screen and stores it in a file called CAS.UPL. Each time you press F9, a new version of CAS.UPL is created and the old version is renamed CAS.BAK.

- F10: This key creates a DARC search string from the structure displayed on the screen and stores it as DARC.UPL. Each time you press F10, the DARC.UPL file is renamed DARC.BAK and a new DARC.UPL is created.

- (SHIFT) F7: This key removes all free sites from the query structure.

- (SHIFT) F9:   The reverse of (SHIFT) F7, this key provides the maximum possible number of free sites at every atom in the query structure.

- (ALT) Fi: The key combination (ALT) Fi, where $i = 3, \ldots, 10$, will cause the generation of an i-membered nonaromatic ring. The new ring will appear on the screen attached to the cursor and can be used in query structure building.

5E3G. Uploading Query Structures

When your query structure is complete in all aspects (remember the free sites!), it can be processed by MOLKICK, which will automatically produce the ROSDAL string that corresponds to it. This is carried out by striking function key F7, which will produce the new ROSDAL string and store it, overwriting any older ROSDAL strings that may be in memory. Once this is done, you may leave MOLKICK by pressing ALT-SHFT and returning to DOS. Now you can start your communications program and connect and log on to DIALOG.

When you have logged on to DIALOG and begun file 390, enter the QS command. This starts the DIALOG structure search program, which prompts you for a ROSDAL string. At this point, press ALT-R and you will see the ROSDAL string appear on your screen. It will be terminated by @@ (appended automatically by MOLKICK), and the search will be carried out. The number of retrievals will be reported to you as usual, and a set will be created containing those retrievals. This set is like any other set created from file 390 by DIALOG, and it can be used in just the same way.

The results from a structure search may suggest a modification that should be made to the query structure in order that a more useful search can be carried out. In this case, just press ALT-SHFT. This will bring back the MOLKICK screen, with the current query structure. You can modify it as you wish and then repeat the upload–search sequence (F7, ALT-SHFT, QS, ALT-R) as before to carry out the second search.

5E3H.   Use of a Database Structure as the Query Structure

It is possible, instead of building a query structure manually or with MOLKICK, to extract a structure from the database and use it as the query structure. This will take the stored ROSDAL string and resubmit it for a full structure search. In this way, one can retrieve from the database all structures whose ROSDAL strings are the same as the query structure. This will include the original compound, along with all isotopic modifications, stereoisomers, and charged forms:

```
?qs/104663

QS/QC Query Structure Version 1.01
DASD 2FF   DETACHED
```

```
Processing - Atom-by-atom match started
      S1        2  QS / 104663 ID  BN=104663 CN=2-bromo-thiophene
```

?t/gs bi/1-2

```
 1/GSBI/1
1422524
  Lawson No: 16857
  Beilstein Cit: 5-17
  Molecular Formula: C4H2BrDS
  Linear search formula: C4H2DBrS
  Molecular Weight: 164.04
  Synonym: 2-Deuterio-5-bromthiophen
  No. of Ref: 6
Graphic Structure:
'1422524'
```

```
  Data Present:
   Data  Ref
   +Ref/Only UDF    Data Type
       /5      PR Preparative Data
       /1      CR Chemical Reactions
      4/3      PP Physical Properties
       /2      KW Short File Keywords
```

```
 1/GSBI/2
104663
2-bromo-thiophene
  German Chem. Name: 2-Brom-thiophen
  Lawson No: 16857
  Beilstein Cit: 4-17-00-00245; 0-17-00-00033; 2-17-00-00036;
1-17-00-00018;
               5-17
  Molecular Formula: C4H3BrS
  Molecular Weight: 163.03
  No. of Ref: 130
Graphic Structure:
'104663'
```

```
Data Present:
 Data  Ref
 +Ref/Only UDF    Data Type
    10/1     PR Preparative Data
    33/11    CR Chemical Reactions
    60/25    PP Physical Properties
     5/      DR Characterization Derivative
      /8     KW Short File Keywords
```

In a second example, one of the enantiomeric forms of glucose is used as the query structure and 48 like structures are retrieved. The first 6 of these are displayed, and it can be seen that these 48 compounds are either stereoisomers or enantiomers of glucose or isotopically labeled forms of any of these stereoisomers:

```
?s glucose/cn and mf=c6h12o6

           607  GLUCOSE/CN  (see also 3 related terms)
           293  MF=C6H12O6
     S2      3  GLUCOSE/CN AND MF=C6H12O6

?t /bn/3

 2/BN/3
1724615

?qs/1724615

QS/QC Query Structure Version 1.01
DASD 2FF  DETACHED
Processing - Atom-by-atom match started
     S3     48  QS / 1724615 ID  BN=1724615 CN=D-Glucose

?t /cn sy gs/1-6

 3/CNSYGS/1
  Synonym: (1-(14)C)D-Galaktose
Graphic Structure:
'3051236'
```

```
 3/CNSYGS/2
  Synonym: Glucose
```

Graphic Structure:
'2413681'

3/CNSYGS/3
  Synonym: Glucoseradikal
Graphic Structure:
'2365787'

3/CNSYGS/4
  Synonym: Galactose
Graphic Structure:
'2327208'

3/CNSYGS/5
  Synonym: Mannose
Graphic Structure:
'2327207'

3/CNSYGS/6
  Synonym: D-Glycose-1-(13C)
Graphic Structure:
'2263682'

5E4. Graphics Output

All output from answer sets derived in DIALOG can be DISPLAYed or TYPEd at your terminal, as is described in Chapter 4 of this manual. Alternatively, such results can be PRINTed, as is explained in section 2B12, in which case the material will be printed on central printers in Palo Alto and mailed to you.

When the data to be output include graphics structures, the PRINT command should not be used. If it is used, all that will appear in place of the graphics structure is the ROSDAL string for the compound. If the DISPLAY or TYPE command is used with such output, the resulting files will also contain the ROSDAL strings, but these can be captured and subsequently printed with the help of a local program which converts a ROSDAL string to a structure diagram. This local program is called GEOFF and is described in the next section.

5E4A. Printing Structures: Graphic Enhanced Output File Format (GEOFF)

GEOFF is a DIALOG program which can be loaded on your PC and used to print files which contain ROSDAL strings. When it encounters a ROSDAL string in an output file, GEOFF produces the corresponding chemical structure diagram in a format suitable for your printer to manage.

The programs that constitute GEOFF are provided by DIALOG on a distribution disk and should all be copied into a unique directory, e.g. \GEOFF, on your hard disk. One of the files within GEOFF is called READ.ME and contains detailed information concerning operation of the program.

To use GEOFF to print a file, the following command string should be typed:

$$\text{GEOFF } \langle \text{arg-1} \rangle \ \langle \text{arg-2} \rangle \ \langle \text{arg-3} \rangle \ \langle \text{arg-4} \rangle$$

where the arguments are as follows:

- arg-1: Name of the output file, with directory path.
- arg-2: Printer type and printer density (see below).
- arg-3: Data file path.
- arg-4: Display options controlling display of mass, charge, valence, and radical information.

The printer type and print density should be selected from the following:

- Epson and IBM-compatible:

EPSD	72×60 dots per inch (dpi)
EPDD	72×120 dpi

- HP Think, Quiet, Desk Jet compatibles:

TJ96	96 × 96 dpi
TJMD	192 × 96 dpi
TJ192	192 × 192 dpi

- HP Laser Jet compatibles:

LJ75	75 × 75 dpi
LJ100	100 × 100 dpi
LJ150	150 × 150 dpi
LJ300	300 × 300 dpi

If your output file is called OUTPUT.NEW and is in the directory \PCPLOT, then it can be printed with all options in highest quality on a laser jet printer with the command

```
C:\GEOFF\GEOFF \PCPLOT\OUTPUT.NEW LJ300 SCMRV
```

Once the correct command sequence is established, the output can be varied by varying the last argument, the SCMRV. An S in this argument will eliminate all output except the structure diagram. C, M, R, and V will produce any charge, mass, radical, or valency data associated with the structure to be printed.

The sample sequence given below shows how a file can be created on the DIALOG mainframe, typed (including a ROSDAL string), and then downloaded and typed again by GEOFF, which converts the ROSDAL string to a structure diagram, with and without various additional data such as mass and charge specifications.

First the file is created by means of some retrieval option on DIALOG:

```
?s pyridine/cn and mf=c5h5n
          9452  PYRIDINE/CN  (see also Py, pyridin)
            26  MF=C5H5N
     S4      1  MF=C5H5N AND PYRIDINE/CN
```

Then an output file is generated containing all the required data. The BN of this compound is extracted:

```
?t/bn

 4/BN/1
103233
```

and used in a **QS** command:

```
?qs/103233


QS/QC Query Structure Version 1.01
DASD 2FF   DETACHED
Processing - Atom-by-atom match started
     S5       43  QS / 103233 ID  BN=103233 CN=pyridine
```

These 43 compounds will all have the same structure as pyridine, but will also have charges and isotopes imbedded in that structure. A display in format 19 (full record) of the second entry in the file is requested, via a `type` command:

```
?t/19/2

 5/19/2
1637469
   Lawson No: 24225
   Beilstein Cit: 5-20
   Molecular Formula: C5H3D2N
   Linear search formula: (C5H3D2N)+
   Molecular Weight: 81.11
   No. of Charges: 1
   No. of Ref: 1
Graphic Structure:
'1637469'
(CO"BRN=1637469")1N(X504,Y575,V4,U1,+1),2(X254,Y432),3(X752,
Y432),4(X254,Y144),5(X752,Y144),6(X504,Y0),7D(X0,Y594),8D(X999,Y5
94)
,2-1=3,4=2-7,5-3-8,6-4,6=5.'
   Data Present:
    Data  Ref
    +Ref/Only UDF    Data Type
        /1    PR Preparative Data
        /1    KW Short File Keywords
Preparative Data
   Preparation (Ref. 1)
   Refs.
      1, Shida, Kato, CHPLBC, Chem.Phys.Lett., 68(1979)106,108
Data of Short File
   ESR (Ref. 1)
   Refs.
      1, Shida, Kato, CHPLBC, Chem.Phys.Lett., 68(1979)106,108
Refs.
   1, Shida, Kato, CHPLBC, Chem.Phys.Lett., 68(1979)106,108
```

and this information, ROSDAL string and all, is captured in a download file called
`\PCPLOT=DIAL12`.

At this stage, the DIALOG session was terminated with a logoff command and the
PC was used in local mode. The program GEOFF was activated with different argument
strings, to produce the following displays:

```
C:>\GEOFF\GEOFF  \PCPLOT\DIAL12 LJ300 \GEOFF  SCMRV
```

```
C:>\GEOFF\GEOFF  \PCPLOT\DIAL12 LJ300 \GEOFF  N

 1637469/19
1637469
  Lawson No: 24225
  Beilstein Cit: 5-20
  Molecular Formula: C5H3D2N
  Linear search formula: (C5H3D2N)+
  Molecular Weight: 81.11
  No. of Charges: 1
  No. of Ref: 1
Graphic Structure:
'1637469'
```

```
  Data Present:
   Data  Ref
   +Ref/Only UDF    Data Type
      /1    PR Preparative Data
      /1    KW Short File Keywords
Preparative Data
  Preparation (Ref. 1)
  Refs.
     1, Shida, Kato, CHPLBC, Chem.Phys.Lett., 68(1979)106,108
Data of Short File
  ESR (Ref. 1)
```

```
    Refs.
       1, Shida, Kato, CHPLBC, Chem.Phys.Lett., 68(1979)106,108
Refs.
    1, Shida, Kato, CHPLBC, Chem.Phys.Lett., 68(1979)106,108


C:>\GEOFF\GEOFF  \PCPLOT\DIAL12 LJ75 \GEOFF  SCM
'1637469'
```

VI. Sample Search Sessions in Beilstein Online

6A. INTRODUCTION AND PURPOSE

This chapter is designed to provide the user with examples of searches in Beilstein Online. While numerous examples of search procedures, including login and logoff, have been shown elsewhere in this manual, this is the only location in the manual where examples are presented as complete sessions.

6B. SAMPLE STRUCTURE AND DATA SEARCH

This first example begins at the first connection to the Telenet network and ends at logoff after the search is completed.

The example search is for all aziridine ring (three membered ring containing one nitrogen) compounds in the database with a molecular weight between 150 and 200 and for which there is reported an infrared spectrum. The query structure is assembled using MOLKICK and is then transferred to DIALOG as a QS. The search is completed, giving 5047 retrievals.

A molecular weight search for compounds with a molecular weight between 150 and 200 is too broad and exceeds DIALOG's system limits. A message to this effect is given, and the search is terminated at MW = 177.95. This problem can be overcome by doing three MW searches and combining the results with a logical OR.

All compounds in the database for which an infrared spectrum has been reported are retrieved with a DP= search.

Finally, the four compounds which satisfy all three search criteria are isolated by combining the three results files already obtained in a logical AND, and the records are displayed using GEOFF, which allows display of the structures.

```
Welcome to DIALOG
Dialog level 22.01.8D

Last  logoff  22dec89 08:41:56
*****************************************************************

As of 09:18 EST on 12/23/89 you had 3 DIALMAIL message(s)
waiting.
*****************************************************************

Logon file204 23dec89 09:18:19
*** File 060 is not working ***
*** File 480 is not working ***
COPR. (c) DIALOG INFORMATION SERVICES, INC. ALL RIGHTS RESERVED.
NO CLAIM TO ORIG. U.S. GOVT. WORKS.
```

```
    *** The DIALOG computers will be operating all day on both ***
    ***            December 25th and January 1st.              ***


Announcements:
   New: D&B-EUROPEAN DUN'S MARKET IDENTIFIERS (File 521)
   Duplicate Detection Available!  (see HOMEBASE for details)
   New: SAN JOSE MERCURY NEWS (File 634)
   New: PETROLEUM EXPLORATION & PRODUCTION (File 987)
   New: ARAB INFORMATION BANK (File 465)
   Reloaded: ICC BRITISH COMPANY DIRECTORY (File 561)

    >>> Enter BEGIN HOMEBASE for Dialog Announcements <<<
    >>>     of new databases, price changes, etc.      <<<
    >>>        Announcements last updated 18dec89       <<<


File 204:ONTAP CA SEARCH (99,122 documenTs)
       (Copr. Amer. Chem. Soc.)

     Set  Items  Description
     ---  -----  -----------
? b 390


     23dec89 09:19:09 User300832 Session D244.1
          $0.23    0.015 Hrs File204
   $0.23  Estimated cost File204
   $0.23  Estimated cost this search
   $0.23  Estimated total session cost   0.015 Hrs.


File 390:Beilstein ONLINE - 3,054,888 compounds - Nov. 1989


     Set  Items  Description
     ---  -----  -----------

?qs

QS/QC Query Structure Version 1.01
Enter ROSDAL connection table
Line No.  1 (enter '.a' to end or 'ABORT' to quit)
```

(Here, the query structure that was generated in MOLKICK is passed from your PC memory through telecommunications to the DIALOG processor. After each line of data, the processor prompts for another line until the terminator—the аа, which MOLKICK appends to the string—is detected. Then the search is carried out automatically.)

```
? 1*1-4*3,1-3N*1-2*2-1.
Line No.  2 (enter '.@' to end or 'ABORT' to quit)
? @@

DASD 2FF  DETACHED
Processing - Atom-by-atom match started
      S1    5047  QS  ID Rosdal: 1*1-4*3,1-3N*1-2*2-1
```

The MW search cannot be done in a single step, because it exceeds system processing limits:

```
?s mw=150:200
  >>>File 390 processing for MW = 200 : MW = 150 started at MW=
   150.00 stopped at MW=       177.95
      S2  168222  MW=150:170
```

A simple way around this is to break the search into more manageable segments, linked together with OR:

```
?s mw=150:170 or mw=170:190 or mw=190:200
Processing
Processing
Processing
          113586  MW=150 : MW=170
          148677  MW=170 : MW=190
           87831  MW=190 : MW=200
     S3   350040  MW=150:170 OR MW=170:190 OR MW=190:200
```

The retrieval of compounds having infrared spectra is straightforward:

```
?s dp=irs
     S4   10096  DP=IRS  (Infrared Spectrum Wavenumber)
```

A Boolean AND combination of these three results files leads to the four compounds which meet the specified criteria:

```
?c 1 and 3 and 4
            5047  1
          350040  3
           10096  4
     S5       4  1 AND 3 AND 4
```

This file can be TYPEd or DISPLAYed using a format which includes the basic information, the structure, and a KWIC term, which will produce the molecular weight and infrared spectral data. If this is done online, the atomic coordinates and the ROSDAL string will appear in place of the structure:

```
?t /bi gs k/1

 5/BIGSK/1
148755
1-benzyl-aziridine-2-carboxylic acid methyl ester
   German Chem. Name: 1-Benzyl-aziridin-2-carbonsaeure-methylester
   Lawson No: 26262, 14140, 289
   Beilstein Cit: 4-22-00-00004; 5-22
   Molecular Formula: C11H13NO2
   Molecular Weight: 191.23
   No. of Ref: 7
Graphic Structure:
'148755'
(CO"BRN=148755")1(X529,Y846),2N(X699,Y747),3(X294,Y846),
4(X699,Y944),5(X817,Y543),6O(X176,Y642),7O(X206,Y999),8(X699,Y339
),
9(X0,Y642),10(X503,Y339),11(X798,Y169),12(X405,Y169),13(X699,Y0),
14(X503,Y0)
,2-1-3,4-1,5-2-4,6-3=7,8-5,9-6,10=8-11,12-10,13=11,14=12,14-13.'
   Data Present:
    Data  Ref
    +Ref/Only UDF    Data Type
       1/2    PR Preparative Data
       2/4    CR Chemical Reactions
       8/     PP Physical Properties
       /2     KW Short File Keywords
   Molecular Formula:  C11H13NO2
   Molecular Weight:  191.23
Vibrational Spectra
   IR Spectrum
      Wavenumber : 4000 - 800 cm**-1; Solvent: CCl4; (Ref. 5
      handbook)
```

If, however, the entire file is DISPLAYed in DIALOG and captured using DIALOGLINK, it can subsequently be printed using GEOFF (see section 5E4), and the structures will be displayed in the offline printout:

```
?t /bi gs k/1-4

 5/BIGSK/1
148755
1-benzyl-aziridine-2-carboxylic acid methyl ester
  German Chem. Name: 1-Benzyl-aziridin-2-carbonsaeure-methylester
  Lawson No: 26262, 14140, 289
  Beilstein Cit: 4-22-00-00004; 5-22
  Molecular Formula: C11H13NO2
  Molecular Weight: 191.23
  No. of Ref: 7
Graphic Structure:
'148755'
```

```
  Data Present:
   Data  Ref
   +Ref/Only UDF    Data Type
      1/2      PR Preparative Data
      2/4      CR Chemical Reactions
      8/       PP Physical Properties
       /2      KW Short File Keywords
  Molecular Formula:  C11H13NO2
  Molecular Weight:  191.23
Vibrational Spectra
  IR Spectrum
      Wavenumber : 4000 - 800 cm**-1; Solvent: CCl4; (Ref. 5
      handbook)

 5/BIGSK/2
125240
(1-benzyl-3-methyl-aziridin-2-yl)-methanol
  German Chem. Name: (1-Benzyl-3-methyl-aziridin-2-yl)-methanol
  Lawson No: 24687, 14140
  Beilstein Cit: 4-21-00-00016
  Molecular Formula: C11H15NO
  Molecular Weight: 177.25
  No. of Ref: 2
```

```
Graphic Structure:
'125240'
```

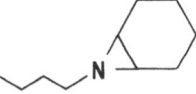

```
   Data Present:
    Data  Ref
    +Ref/Only UDF    Data Type
       1/       PR Preparative Data
       4/       PP Physical Properties
   Molecular Formula:  C11H15NO
   Molecular Weight:  177.25
Vibrational Spectra
   IR Spectrum
       Wavenumber : 3333 - 667 cm**-1; Solvent: CHCl3; (Ref. 2
       handbook)


 5/BIGSK/3
104979
7-butyl-7-aza-norcarane
   German Chem. Name: 7-Butyl-7-aza-norcaran
   Lawson No: 24140, 2844
   Beilstein Cit: 4-20-00-01939; 5-20
   Molecular Formula: C10H19N
   Molecular Weight: 153.27
   No. of Ref: 2
Graphic Structure:
'104979'
```

```
   Data Present:
    Data  Ref
    +Ref/Only UDF    Data Type
       1/1     PR Preparative Data
       5/       PP Physical Properties
   Molecular Formula:  C10H19N
   Molecular Weight:  153.27
Vibrational Spectra
   IR Spectrum
       Wavenumber : 5000 - 667 cm**-1; (Ref. 1 handbook)
```

```
 5/BIGSK/4
82734
cis-2,3-diphenyl-aziridine
  German Chem. Name: cis-2,3-Diphenyl-aziridin
  Lawson No: 24456
  Beilstein Cit: 4-20-00-03901; 5-20
  Molecular Formula: C14H13N
  Molecular Weight: 195.26
  No. of Ref: 37
Graphic Structure:
'82734'
```

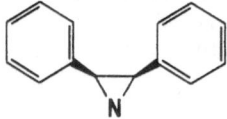

```
  Data Present:
   Data   Ref
   +Ref/Only UDF    Data Type
      3/28    PR Preparative Data
      1/5     CR Chemical Reactions
     13/4     PP Physical Properties
      2/      DR Characterization Derivative
       /3     KW Short File Keywords

  Molecular Formula:  C14H13N
  Molecular Weight:  195.26
Vibrational Spectra
  IR Spectrum
     Wavenumber : 4000 - 833 cm**-1; Solvent: CCl4; (Ref. 28
     handbook)


?logoff

        23dec89 11:47:22 User300832 Session D245.2
            $19.80    0.090 Hrs File390
                $5.45  1 Type(s) in Format 21 (UDF)
               $19.20  4 Type(s) in Format 31 (UDF)
            $24.65  5 Types
               $49.00  1 QS/QC Substructure Search
            $49.00  1 QS/QC Report Total
    $93.45  Estimated cost File390
    $93.45  Estimated cost this search
    $93.54  Estimated total session cost   0.097 Hrs.
Logoff: level 22.01.8 D  11:47:23
```

6C. SAMPLE DATA SEARCH

A compound isolated as crystals with a melting point of 200–203°C gave a UV spectrum maximum at 292 nm in dioxane. It is possible to find compounds with the same properties in the Beilstein Database, and these represent possible structures for the unknown compound.

   Note the use of operators here. The (S) operator is used to ensure that the UV data were measured in dioxane (the operator requires that the dioxane and the UV data be in the same line), and the AND operator requires that the melting point and the UV/solvent data belong to the same record:

```
?s eam=292(s)dioxane and mp=200:203

           340  EAM=292
         21312  DIOXANE  (see also dioxan)
             5  EAM=292(S)DIOXANE
         46965  MP=200 : MP=203
     S6      1  EAM=292(S)DIOXANE AND MP=200:203

?d /bi gs k
         Display 6/BIGSK/1

383659
indoline-2,3-dione
  German Chem. Name: Indolin-2,3-dion
  Lawson No: 25776
  Beilstein Cit: 0-21-00-00432; 4-21-00-04981; 1-21-00-00348;
  2-21-00-00327; 5-21
  Molecular Formula: C8H5NO2
  Molecular Weight: 147.13
  No. of Ref: 485
Graphic Structure:
'383659'
```

```
  Data Present:
   Data  Ref

                              -more-
?

      Display 6/BIGSK/1

  +Ref/Only UDF     Data Type
    44/23   PR Preparative Data
```

```
     184/121  CR Chemical Reactions
      67/25   PP Physical Properties
       9/     DR Characterization Derivative
       /8     KW Short File Keywords
Crystals
  Melting Point:  200 - 201  C; (Ref. 66 handbook...
  Melting Point:  199 - 200  C; (Ref. 69...
  Melting Point:  201 - 202  C; (Ref. 70...
  Melting Point:  201  C; (Ref. 71...
  Melting Point:  200 - 202  C; (Ref. 73...
  Melting Point:  199 - 200  C; Solvent: CHCl3; (Ref.
                  75...
  Melting Point:  200 - 201  C; (Ref. 76...
  Melting Point:  201 - 203  C; (Ref. 77...
  Melting Point:  200 - 202  C; (Ref. 78...
  Melting Point:  202 - 203  C; (Ref. 79...
  Melting Point:  199 - 200  C; Solvent: H2O; (Ref.
                  81...
  Melting Point:  201  C; (Ref. 84...
  Melting Point:  199 - 202  C; (Ref. 85...
  Melting Point:  200  C; (Ref. 88...
  Melting Point:  200  C; (Ref. 89)

                                    -more-
?
       Display 6/BIGSK/1

Electronic Spectra
   Absorption Maxima
     Wavelength: 292, 408 nm; Solvent: dioxane; (Ref. 116
     handbook...

                        - end of display -
```

It is interesting to see how this search converges. To some extent this is because of the absence of data, but nevertheless, only five compounds have the correct UV data and only one also has the correct melting point.

6D. IDENTIFICATION OF NATURAL PRODUCTS

During the structure determination of natural products, it is often possible to assign parts of the structure according to the origin of the compound and the structure of etiologically related compounds. Such a partial structure, together with some physical properties data, is often sufficient to support a complete structure assignment.

In the example shown here, an alkaloid derived from a senecio plant was found to have a melting point of 208–209°C and an optical rotation of −92°, measured at the sodium *D* line, in methanol. The compound was presumed to be a pyrrolizidine alkaloid, because these are the only alkaloid types found in plants of the senecio family. This permitted a substructure search for compounds containing the pyrrolizidine skeleton. The query structure was generated in MOLKICK and then uploaded into a QS:

```
Welcome to DIALOG
Dialog level 22.01.8D


Last  logoff  28dec89 16:50:43
Reconnected in file 390  28dec89 17:20:44


File 390:Beilstein ONLINE - 3,054,888 compounds - Nov. 1989

      Set  Items  Description
      ---  -----  -----------
?qs


QS/QC Query Structure Version 1.01
Enter ROSDAL connection table
Line No.  1 (enter '.@' to end or 'ABORT' to quit)
```

(Here, the query structure that was generated in MOLKICK is passed from your PC memory through telecommunications to the DIALOG processor. After each line of data, the processor prompts for another line until the terminator—the @@, which MOLKICK appends to the string—is detected. Then the search is carried out automatically.)

```
? 1--3*2-4*2-5N-1-6-9-100*1,2-110*1,5-8*2-7*2-6.
Line No.  2 (enter '.@' to end or 'ABORT' to quit)
? @@


DASD 2FF  DETACHED
Processing - Atom-by-atom match started
      S7      96  QS  ID Rosdal: 1--3*2-4*2-5N-1-6-9-100*1,
                                  2-110*1...
```

The set (S7) of 96 hits was then combined with compounds having the correct melting point and optical rotatory power, and in this way, just two retrievals resulted:

```
?s s7 and mp=205:215 and orp=-100:-90
           96  S7
        96819  MP=205 : MP=215
         3658  ORP=-100 : ORP=-90
   S8       2  S7 AND MP=205:215 AND ORP=-100:-90
```

These two compounds were then displayed:

<u>?t /19/1-2</u>

```
 8/19/1
1089495
   Lawson No: 31888
   Beilstein Cit: 5-27
   Molecular Formula: C18H27N06
   Molecular Weight: 353.41
   Synonym: Rosmarinin
   No. of Ref: 1
Graphic Structure:
'1089495'
```

```
   Data Present:
    Data  Ref
    +Ref/Only UDF    Data Type
       1/      PR Preparative Data
       2/      PP Physical Properties
Preparative Data
   Isolation from Natural Product
       Senecio angulatus L. (Ref. 1)
       Refs.
         1, Porter, Geissman, JOCEAH, J.Org.Chem., 27(1962)4132
Crystals
   Melting Point: 205-208 C; Solvent: ethyl acetate; (Ref. 1)
   Refs.
       1, Porter, Geissman, JOCEAH, J.Org.Chem., 27(1962)4132
Optical Properties
   Optical Rotatory Power: -90.2 deg; Wavelength: 589 nm; Type:
   (alpha); Solvent: ethanol; Concentration: 0.02; Temp: 25 C;
   (Ref. 1)
   Refs.
       1, Porter, Geissman, JOCEAH, J.Org.Chem., 27(1962)4132
Refs.
   1, Porter, Geissman, JOCEAH, J.Org.Chem., 27(1962)4132

 8/19/2
47181
2 alpha,12-dihydroxy-(1 alpha H)-1,2-dihydro-senecionan-11,16-dione
```

```
  German Chem. Name: 2 alpha,12-Dihydroxy-(1 alpha
      H)-1,2-dihydro-senecionan-11,16-dion
  Lawson No: 25180, 2160
  Beilstein Cit: 4-27-00-06661; 5-21
  Molecular Formula: C18H27NO6
  Molecular Weight: 353.41
  No. of Ref: 6
Graphic Structure:
'47181'
```

```
  Data Present:
   Data  Ref
   +Ref/Only UDF    Data Type
      /1     SD Constitutional Data
      2/     PR Preparative Data
      6/     PP Physical Properties
      /1     KW Short File Keywords
Structural Data
   Related Structure
      (Comment: Configuration.) (Ref. 1 handbook)
      Refs.
        1, Warren, v Klemperer, JCSOA9, J.Chem.Soc., 1958 4572
Preparative Data
   Isolation from Natural Product
      Isolierung aus Senecio rosmarinifolius (Ref. 2 handbook)
      Isolierung aus anderen Senecio-Arten (Ref. 3 handbook),
      (Ref. 4 handbook)
      Refs.
        2, de Waal, Onderstepoort J. veterin. Sci., 15 (1940)
           241, 245
        3, Koekemoer, Warren, JCSOA9, J.Chem.Soc., 1951 66
        4, Richardson, Warren, JCSOA9, J.Chem.Soc., 1943 452
Crystals
   Melting Point: 209 C; Solvent: acetone; (Ref. 4 handbook)
   Melting Point: 208 C; Solvent: acetone, ethanol; (Ref. 2
   handbook)
   Refs.
      2, de Waal, Onderstepoort J. veterin. Sci., 15 (1940) 241,
         245
      4, Richardson, Warren, JCSOA9, J.Chem.Soc., 1943 452
Optical Properties
```

```
   Optical Rotatory Power: -91.5 deg; Wavelength: 589 nm; Type:
   (alpha); Solvent: methanol; Concentration: c=1; Temp: 24 C;
   (Ref. 4 handbook)
   Optical Rotatory Power: -94 deg; Wavelength: 589 nm; Type:
   (alpha); Solvent: ethanol; Concentration: c=0.5; Temp: 20 C;
 (Ref. 2 handbook)
   Optical Rotatory Power: -120 deg; Wavelength: 589 nm; Type:
   (alpha); Solvent: CHCl3; Concentration: c=0.6; Temp: 20 C;
(Ref. 2 handbook)
   Refs.
      2, de Waal, Onderstepoort J. veterin. Sci., 15 (1940) 241,
         245
      4, Richardson, Warren, JCSOA9, J.Chem.Soc., 1943 452
Electronic Spectra
   UV/VIS Spectrum
      Wavelength: 200 - 235 nm; Solvent: H2O; (Ref. 5 handbook)
      Refs.
         5, Leisegang, JSACAT, J.S.Afr.Chem.Inst., (N.S.) 3
            (1950) 73, 75
Data of Short File
   Further information (Ref. 6)
   Refs.
      6, Hrbek, et al, CCCCAK, Collect.Czech.Chem.Commun.,
         37(1972)3918,3930,3933
Refs.
   1, Warren, v Klemperer, JCSOA9, J.Chem.Soc., 1958 4572
   2, de Waal, Onderstepoort J. veterin. Sci., 15 (1940) 241, 245
   3, Koekemoer, Warren, JCSOA9, J.Chem.Soc., 1951 66
   4, Richardson, Warren, JCSOA9, J.Chem.Soc., 1943 452
   5, Leisegang, JSACAT, J.S.Afr.Chem.Inst., (N.S.) 3 (1950) 73,
      75
   6, Hrbek, et al, CCCCAK, Collect.Czech.Chem.Commun.,
      37(1972)3918,3930,3933
```

Both of these compounds seem to be a good match for the unknown. They are also a good match for each other, and it seems quite possible that Porter and Geissman's "Rosmarinin" (BN 1089495) is an incompletely characterized form of the other compound (BN 47181), whose stereochemistry is fully described.

6E. IDENTIFICATION FROM PHYSICAL PROPERTIES

An unknown compound was isolated as crystals in a *p*21 space group and with a melting point of 200–205°C. The compound was optically active, with a specific rotation of −11°, measured in ethanol at the sodium *D* line. It showed no evidence of mutarotation. We want to find candidates in the Beilstein Database for this material.

This can be accomplished in one search using ORP, MP, and CSG data. Compounds responding are combined in a Boolean NOT with the 1053 compounds which do show mutarotation. The result is two compounds which have the three observed properties and which are not reported to show mutarotation:

```
?s (orp=-25:-5 and mp=200:205 and csg=p2 1) not dp=mut

        29476   ORP=-25 : ORP=-5
        64810   MP=200 : MP=205
           32   CSG=P2 1
         1053   DP=MUT  (Mutarotation)
    S9      2   (ORP=-25:-5 AND MP=200:205 AND CSG=P2 1) NOT
                DP=MUT


?t /bi gs k/1-2

 9/BIGSK/1
224754
5,6-dihydroxy-4a,8-dimethyl-3-methylene-decahydro-azuleno(6,5-b)
furan-2-one
   German Chem. Name:5,6-Dihydroxy-4a,8-dimethyl-3-methylen-
              decahydro- azuleno(6,5-b)furan-2-on
   Lawson No: 19058
   Beilstein Cit: 4-18-00-01199
   Molecular Formula: C15H2204
   Molecular Weight: 266.34
   No. of Ref: 2
Graphic Structure:
'224754'
```

```
   Data Present:
    Data  Ref
    +Ref/Only UDF    Data Type
       1/        PR Preparative Data
       7/        PP Physical Properties
Crystals
   Melting Point: 202-203 C; Solvent: methanol; (Ref. 2 handbook)
    Space Group
        P2 1  (Ref. 2 handbook)
Optical Properties
   Optical Rotatory Power: -10.7 deg; Wavelength: 589 nm; Type:
   (alpha); Solvent: ethanol; Concentration: c=2; Temp: 29...
```

```
 9/BIGSK/2
102127
(2 R)-3 c,5,7-Triacetoxy-2 r-(3,4,5-triacetoxy-phenyl)-chroman
   German Chem. Name: (2 R)-3 c,5,7-Triacetoxy-2r-(3,4,5-tri-
                  acetoxy- phenyl)-chroman
   Lawson No: 17685, 1155
   Beilstein Cit: 4-17-00-03892; 5-17
   Molecular Formula: C27H26013
   Molecular Weight: 558.49
   No. of Ref: 8
Graphic Structure:
'102127'
```

```
   Data Present:
    Data  Ref
     +Ref/Only UDF    Data Type
        1/1      PR Preparative Data
       13/1      PP Physical Properties
        /1      KW Short File Keywords

Crystals
  Melting Point: 195-200 C; Solvent: methanol; (Ref. 3
  handbook...
   Space Group
       P2 1  (Ref. 7 handbook)
Optical Properties
   Optical Rotatory Power: -14.2 deg; Wavelength: 589 nm; Type:
   (alpha); Solvent: acetone; Temp: 20 C; (Ref. 4...
   Optical Rotatory Power: -14 deg; Wavelength: 589 nm; Type:
(alpha); Solvent: acetone; Concentration: c=0.5; Temp: 21...
```

Clearly, the first compound presents a better fit to the data. Its melting point and optical rotation, as reported, just straddle the experimental points, and it should also be noted that in the case of the second compound, the optical rotation was measured in a different solvent (acetone).

Appendices

Appendix A.   Technical Assistance and Support

Beilstein Institute
Springer-Verlag Inc.
DIALOG Information Services
Newsletters
User Meetings
Workshops
CompuServe Bulletin Board and Online Help

A1. BEILSTEIN INSTITUTE

Dr. Gabrielle Ilchmann
Help Desk
Beilstein Institut für Literatur der Organischen Chemie
Varrentrappstrasse 40-42
Carl-Bosch-Haus
D-6000 Frankfurt/M 90, West Germany

Telephone: (49)-69-7917-311 (Beilstein HELP Desk)
FAX: (49)-69-7917-492 (Beilstein FAX Help Line)
Telex: 416 969 BLSTN D
Dialmail: 25281

A2. SPRINGER-VERLAG INC.

Gertraud Griepke, New Media/Handbooks
Springer-Verlag
Tiergartenstrasse 17
D-6900 Heidelberg 1, West Germany

Telephone: (49)-6221-4874-57
FAX: (49)-6221-487366

Dr. Robert Badger
Manager, Electronic Media
Springer-Verlag New York, Inc.
175 Fifth Avenue
New York NY 10010

Telephone: 212-460-1622
FAX: 212-473-6272

Mr. T. Hirano
Springer-Verlag Tokyo, Inc.
37-3, Hongo 3-chome, Bunkyo-ku
Tokyo 113, Japan

Telephone: (81)-3-812-0331
FAX: (81)-3-812-0719

A3. DIALOG INFORMATION SERVICES

DIALOG Information Services, Inc.
3460 Hillview Avenue
Palo Alto CA 94304

Telephone: (800)-334-2564 (800-3-DIALOG)
Telex: 334499 (DIALOG 8PLA)

Accounts Receivable: (415)-858-3763
Customer Administration: (415)-858-3749
Customer Services: (415)-858-3810
Marketing: (415)-858-3785
Training: (415)-858-3792
President (R. Summit):   (415)-858-3777

Insearch Ltd./DIALOG
Box K16
Haymarket
Sydney, NSW 2000, Australia

Telephone: 79027091

Micromedia Limited
Dialog Information Services Inc.
158 Pearl Street
Toronto, Ontario M5H 1L3, Canada

Telephone: (416)-593-5211
Telex: 06524668

Dialog Information Services Inc.
Learned Information Ltd.
Woodside Hinksey Hill
Oxford OX1 5AU, Great Britain

Telephone: (0) (865)-730-275
Telex: 851837704

Kinokuniya Co. Ltd.
P.O. Box 55 Chitose
Tokyo 156, Japan

Telephone: (03)-439-0123

MASIS Center
Maruzen Co. Ltd.
P.O. Box 5335
Tokyo International 10031, Japan

Telephone: (272)-7211
Telex: 78126630

Data Communications of Korea
12th Floor, Korea Stock Exchange
1-116 Yeoeido-Dong
Yeongdungpo-KU
Seoul, Korea

Telex: 28311

Asesores Especializados en Informacion y Documentacion, S.C.
AEID/DIALOG
2a Cerrada de Romero de Torreros, 49A
03100 Mexico D.F., Mexico

Telephone: (905)-543-7207

A4. NEWSLETTERS

There are a number of newsletters about the Beilstein Database and Beilstein activities which are distributed regularly by the various organizations associated with the database. The following is a list of such newsletters and the address to contact should you wish to subscribe. At present, all of these newsletters are distributed without charge.

A4A. Beilstein Brief

The *Beilstein Brief* newsletter is published two or three times per year by the Beilstein Institute. The newsletter describes the many activities of the Beilstein Institute, including the computerization of the Beilstein Database, and the Beilstein Online system. To be placed on the mailing list, contact the nearest Springer-Verlag office, or the Beilstein Institute.

A4B. CHRONOLOG

CHRONOLOG is a monthly newsletter published by Dialog Information Services Inc. It is distributed free of charge to all registered users of the DIALOG system. The newsletter frequently contains informative articles about search technique and system features. Articles may also include instructions for annotating database bluesheets, and new bluesheets are distributed with *CHRONOLOG*.

A5. USER MEETINGS

The Beilstein Institute and Springer-Verlag hold regular meetings where Beilstein Online users can learn and hear about the latest information concerning the system and discuss their experiences, triumphs, and problems. Details on user meetings are published in the *Beilstein Brief* and can also be obtained from Springer-Verlag or from the Beilstein Institute.

A6. WORKSHOPS

The Beilstein Institute, Springer-Verlag, and Dialog Information Services all hold regular training workshops where the Beilstein Online system is taught to new and experienced users. Those attending these sessions get the latest information concerning the system. Details on training workshops can be obtained from customer information mailing, *CHRONOLOG*, the *Beilstein Brief*, the Beilstein Institute, and Springer-Verlag.

A7. COMPUSERVE BEILSTEIN BULLETIN BOARD AND ONLINE HELP

There is an online Beilstein Bulletin Board and Help Forum system available without charge to all users of Beilstein Online. For details of how to access this online system on the CompuServe computer system, contact any of the user assistance offices listed in sections A1–A3 of this appendix.

Appendix B.   Elemental Maximum Valences and Mass Ranges

Appendix B   Elemental Maximum Valences and Mass Ranges

The following is a list of the elements, sorted by their atomic symbols. The list includes their IUPAC names, their maximum valences, and their nominal mass ranges, for use in the Beilstein Online Substructure Search System.

ELEMENT SYMBOL	ELEMENT NAME	VALENCE	MASS RANGE
Ac	Actinium	3	216–231
Ag	Silver	1	100–117
Al	Aluminum	3	23– 30
Am	Americium	3	237–246
Ar	Argon	0	33– 42
As	Arsenic	3	68– 85
At	Astatine	1	198–219
Au	Gold	1	177–203
B	Boron	3	8– 13
Ba	Barium	2	123–144
Be	Beryllium	2	6– 12
Bi	Bismuth	3	196–215
Bk	Berkelium	3	243–251
Br	Bromine	1	74– 90
C	Carbon	4	9– 16
Ca	Calcium	2	37– 50
Cd	Cadmium	2	101–121
Ce	Cerium	3	129–148
Cf	Californium	3	244–254
Cl	Chlorine	1	37– 50
Cm	Curium	3	238–252
Co	Cobalt	2	54– 64
Cr	Chromium	6	46– 56
Cs	Cesium	1	123–144
Cu	Copper	2	58– 68
Dy	Dysprosium	3	149–167
Er	Erbium	3	152–172
Es	Einsteinium	3	242–256
Eu	Europium	3	143–160
F	Fluorine	1	16– 22
Fe	Iron	3	52– 61
Fm	Fermium	3	248–258
Fr	Francium	1	204–224
Ga	Gallium	3	63– 76
Gd	Gadolinium	3	145–162
Ge	Germanium	4	65– 78

Appendix B   Elemental Maximum Valences and Mass Ranges

H	Hydrogen	1	1– 5
He	Helium	0	3– 8
Hf	Hafnium	4	157–183
Hg	Mercury	2	185–206
Ho	Holmium	3	150–170
I	Iodine	1	117–139
In	Indium	3	106–124
Ir	Iridium	2	192–198
K	Potassium	1	37– 47
Kr	Krypton	0	74– 95
La	Lanthanum	3	124–144
Li	Lithium	1	5– 9
Lr	Lawrencium	3	256–257
Lu	Lutetium	3	15–180
Md	Mendelevium	3	255–257
Mg	Magnesium	2	21– 28
Mn	Manganese	7	50– 58
Mo	Molybdenum	6	88–105
N	Nitrogen	5	12– 18
Na	Sodium	1	20– 26
Nb	Niobium	5	88–105
Nd	Neodymium	3	137–151
Ne	Neon	0	17– 24
Ni	Nickel	2	56– 67
No	Nobelium	3	254–256
Np	Neptunium	5	227–241
O	Oxygen	2	13– 20
Os	Osmium	2	180–195
P	Phosphorus	5	28– 34
Pa	Protoactinum	5	225–237
Pb	Lead	4	194–214
Pd	Palladium	2	98–115
Pm	Promethium	3	140–154
Po	Polonium	2	192–218
Pr	Praseodymium	3	134–149
Pt	Platinum	2	173–201
Pu	Plutonium	4	232–246
Ra	Radium	2	213–230
Rb	Rubidium	1	79– 95
Re	Rhenium	7	177–192

Appendix B Elemental Maximum Valences and Mass Ranges

Rh	Rhodium	2	96–110
Rn	Radon	0	204–224
Ru	Ruthenium	2	93–108
S	Sulfur	6	29– 38
Sb	Antimony	5	112–135
Sc	Scandium	3	40– 51
Se	Selenium	2	70– 87
Si	Silicon	4	25– 32
Sm	Samarium	3	141–158
Sn	Tin	4	108–132
Sr	Strontium	2	80– 95
Ta	Tantalum	5	172–186
Tb	Terbium	3	147–164
Tc	Technetium	7	92–107
Te	Tellurium	2	107–135
Th	Thorium	4	223–235
Ti	Titanium	4	41– 52
Tl	Thallium	1	191–210
Tm	Thulium	3	153–176
U	Uranium	6	227–240
V	Vanadium	5	45– 54
W	Tungsten	6	174–189
Xe	Xenon	0	118–144
Y	Yttrium	3	82– 97
Yb	Ytterbium	3	154–177
Zn	Zinc	2	60– 72
Zr	Zirconium	4	81– 99

Appendix C.   Stopwords in the Beilstein Database

Appendix C   Stopwords in the Beilstein Database

The following words are not allowed for searching in the Beilstein Database, as they either are reserved for searching operators (e.g. **AND**) or are so common as to be useless for searching:

> AN
> AND
> BY
> FOR
> FROM
> OF
> THE
> TO
> WITH

Appendix D. Tabulation of Occurrences of Data Fields

Appendix D   Tabulation of Occurrences of Data Fields

Occurrences	Term
1909	ADSP (Adsorption)
1909	ADSPC (Adsorption Comments/refs.)
0	ADSPFD (Adsorption Further Details)
416	ADSPPA (Adsorption Partner)
17	ADSPPRES (Adsorption Pressure)
251	ADSPSOLV (Adsorption Solvent)
146	ADSPTEMP (Adsorption Temperature)
5220	ASSN (Association)
5220	ASSNC (Association Comments/refs.)
0	ASSNFD (Association Further Details)
575	ASSNPA (Association Partner)
5	ASSNPRES (Association Pressure)
318	ASSNSOLV (Association Solvent)
96	ASSNTEMP (Association Temperature)
929	AZ (Azeotrope Components)
1085	AZC (Azeotrope Comments/refs.)
0	AZCONC (Azeotrope Concentrations)
465	AZPRES (Azeotrope Pressure)
543	AZTEMP (Azeotrope Temperature)
8176	BF (Biological Function)
8176	BFC (Biological Function Comments/refs.)
0	BM (Bond Moment)
110	BMC (Bond Moment Comments/refs.)
0	BMTYPE (Bond Moment Type)
471532	BP (Boiling Point)
473828	BPC (Boiling Point Comments/refs.)
0	BPDEC (Boiling Point Decomposition)
471562	BPPRES (Boiling Point Pressure)
97573	CCOL (Color or Other Properties)
114670	CCOLC (Color or Other Properties Comments/refs.)
43	CDIC (Circular Dichroism)
4851	CDICC (Circular Dichroism Comments/refs.)
50	CDICSOLV (Circular Dichroism Solvent)
0	CDN (Density of the Crystal)
2805	CDNC (Density of the Crystal Comments/refs.)
0	CEXP (Coefficient of Expansion)
313	CEXPC (Coefficient of Expansion Comments/refs.)
0	CEXPTEMP (Coefficient of Expansion Temperature)
0	CFR (Cross File Reference)
18550	CH (Number of Charges)
68	CLPAA (Unit Cell Dimension Angle Alpha)
378	CLPAB (Unit Cell Dimension Angle Beta)

Appendix D   Tabulation of Occurrences of Data Fields

```
    69   CLPAC (Unit Cell Dimension Angle Gamma)
  2302   CLPC (Unit Cell Dimension Comments/refs.)
  5688   CLPLA (Unit Cell Dimension Length A)
   638   CLPLB (Unit Cell Dimension Length B)
   664   CLPLC (Unit Cell Dimension Length C)
   622   CLPN (Unit Cell Dimension N)
     0   CMC (Critical Micelle Concentration)
   166   CMCC (Critical Micelle Concentration Comments/refs.)
    21   CMCSOLV (Critical Micelle Concentration Solvent)
     8   CMCTEMP (Critical Micelle Concentration Temperature)
509014   CN (Chemical Name)
   413   CP (Heat Capacity Cp)
   959   CPC (Heat Capacity Cp Comments/refs.)
   403   CPTEMP (Heat Capacity Cp Temperature)
   277   CPTP (Transition Point Crystalline Modification)
   843   CPTPC (Transition Point Crystal. Mod. comments/refs.)
   247   CP0 (Heat Capacity Cp0)
   454   CP0C (Heat Capacity Cp0 Comments/refs.)
   258   CP0TEMP (Heat Capacity Cp0 Temperature)
    92   CRDN (Critical Density)
   156   CRDNC (Critical Density Comments/refs.)
   154   CRPRES (Critical Pressure)
   247   CRPRESC (Critical Pressure Comments/refs.)
   196   CRTEMP (Critical Temperature)
   428   CRTEMPC (Critical Temperature Comments/refs.)
     0   CRVOL (Critical Volume)
   109   CRVOLC (Critical Volume Comments/refs.)
   590   CSG (Space Group)
  1783   CSGC (Space Group Comments/refs.)
     0   CSOL (Crystallization Solvent)
     0   CSOLC (Crystallization Solvent Comments/refs.)
     0   CSOLMOL (Crystallization Mols of Solvent)
  1123   CSYS (Crystal System)
  1702   CSYSC (Crystal System Comments/refs.)
    18   CV (Heat Capacity Cv)
    99   CVC (Heat Capacity Cv Comments/refs.)
    21   CVTEMP (Heat Capacity Cv Temperature)
     0   DECOMP (Decomposition)
  4776   DECOMPC (Decomposition Comments/refs.)
     0   DECOMPS (Decomposition Solvent)
   578   DIC (Dielectric Constant)
  2571   DICC (Dielectric Constant Comments/refs.)
   337   DICF (Dielectric Constant Frequency)
   697   DICTEMP (Dielectric Constant Temperature)
```

274	DISC (Static Dielectric Constant)
393	DISCC (Static Dielectric Constant Comments/refs.)
282	DISCTEMP (Static Dielectric Constant Temperature)
2416	DM (Dipole Moment)
11407	DMC (Dipole Moment Comments/refs.)
2305	DMMETHOD (Dipole Moment Method)
2200	DMSOLV (Dipole Moment Solvent)
773	DMTEMP (Dipole Moment Temperature)
130946	DN (Density)
131323	DNC (Density Comments/refs.)
130958	DNMSRT (Density Measurement Temperature)
116971	DNREFT (Density Reference Temperature)
75297	DR (Derivative)
75308	DRC (Derivative Comments/refs.)
5798	DX (Dissociation Exponent)
38483	DXC (Dissociation Exponent Comments/refs.)
743	DXGROUP (Dissociation Exponent Group)
4686	DXMETHOD (Dissociation Exponent Method)
5203	DXSOLV (Dissociation Exponent Solvent)
4024	DXTEMP (Dissociation Exponent Temperature)
5844	DXTYPE (Dissociation Exponent Type)
22018	EAM (Absorption Maxima)
94623	EAMC (Absorption Maxima Comments/refs.)
0	EAMEAC (Absorption Maxima Extinction/Absorption Coefft.)
20858	EAMSOLV (Absorption Maxima Solvent)
14611	EAS (Ultraviolet/Visible Spectrum Wavelength)
42760	EASC (Ultraviolet/Visible Spectrum Comments/refs.)
14760	EASSOLV (Ultraviolet/Visible Spectrum Solvent)
0	EBC (Energy Barriers)
2686	EBCC (Energy Barriers Comments/refs.)
0	EBCTYPE (Energy Barriers Bond Type)
12	ED (Ecological Data)
12	EDC (Ecological Data Comments/refs.)
32	EDIS (Dissociation Energy)
723	EDISC (Dissociation Energy Comments/refs.)
175	EDISTYPE (Dissociation Energy Bond Type)
0	EZP (Zero Point Energy)
65	EZPC (Zero Point Energy Comments/refs.)
363	FLM (Fluorescence Maxima)
578	FLMC (Fluorescence Maxima Comments/refs.)
411	FLMSOLV (Fluorescence Maxima Solvent)
243	FLS (Fluorescence Spectrum Wavelength)
357	FLSC (Fluorescence Spectrum Comments/refs.)
0	FLSMETH (Fluorescence Spectrum Excitation Method)

Appendix D   Tabulation of Occurrences of Data Fields

241	FLSSOLV (Fluorescence Spectrum Solvent)
461024	GC (German Chemical Name)
30251	GCOM (General Comment)
33	GFOR (Gibbs Energy of Formation)
510	GFORC (Gibbs Energy of Formation Comments/refs.)
0	GFORPRES (Gibbs Energy of Formation Pressure)
24	GFORTEMP (Gibbs Energy of Formation Temperature)
725	HCOM (Enthalpy of Combustion)
1706	HCOMC (Enthalpy of Combustion Comments/refs.)
15	HCOMPRES (Enthalpy of Combustion Pressure)
382	HCOMTEMP (Enthalpy of Combustion Temperature)
0	HDIS (Enthalpy of Dissociation (Electrolytic))
203	HDISC (Enthalpy of Dissociation Comments/refs.)
27	HDISTEMP (Enthalpy of Dissociation Temperature)
223	HFOR (Enthalpy of Formation)
2858	HFORC (Enthalpy of Formation Comments/refs.)
3	HFORPRES (Enthalpy of Formation Pressure)
72	HFORTEMP (Enthalpy of Formation Temperature)
10	HHYD (Enthalpy of Hydrogenation)
138	HHYDC (Enthalpy of Hydrogenation Comments/refs.)
7	HHYDPROD (Enthalpy of Hydrogenation Product Name)
12	HHYDTEMP (Enthalpy of Hydrogenation Temperature)
379	HMP (Enthalpy of Melting)
958	HMPC (Enthalpy of Melting Comments/refs.)
26	HPTP (Other Phase Transition Enthalpies)
365	HPTPC (Other Phase Transition Enthalpies Comments/refs.)
88	HSUB (Enthalpy of Sublimation)
297	HSUBC (Enthalpy of Sublimation Comments/refs.)
16	HSUBTEMP (Enthalpy of Sublimation Temperature)
961	HVP (Enthalpy of Vaporization)
1917	HVPC (Enthalpy of Vaporization Comments/refs.)
105	HVPPRES (Enthalpy of Vaporization Pressure)
417	HVPTEMP (Enthalpy of Vaporization Temperature)
73	IEP (Isoelectric Point PH)
109	IEPC (Isoelectric Point Comments/refs.)
40	IEPSOLV (Isoelectric Point Solvent)
125	IP (Ionization Potential)
2913	IPC (Ionization Potential Comments/refs.)
140	IPMETHOD (Ionization Potential Method)
6055	IRM (Infrared Bands Wavelength)
46393	IRMC (Infrared Bands Comments/refs.)
4409	IRMSOLV (Infrared Bands Solvent)
10096	IRS (Infrared Spectrum Wavenumber)
32798	IRSC (Infrared Spectrum Comments/refs.)

Appendix D   Tabulation of Occurrences of Data Fields

```
 6534   IRSSOLV (Infrared Spectrum Solvent)
34179   IS (Isolation from Natural Product)
34179   ISC (Isolation from Natural Product Comments/refs.)
 4859   LL (Liquid/Liquid Systems)
 4859   LLC (Liquid/Liquid Systems Comments/refs.)
    0   LLFD (Liquid/Liquid Systems Further Details)
  455   LLPA (Liquid/Liquid Systems Partner)
   49   LLPRES (Liquid/Liquid Systems Pressure)
  798   LLSOLV (Liquid/Liquid Systems Solvent)
  608   LLTEMP (Liquid/Liquid Systems Temperature)
   25   LPTP (Transition Points of Liquid Modification)
 3416   LPTPC (Transition Points of Liquid Mod. Comments/refs.)
 2540   LS (Liquid/Solid Systems)
 2540   LSC (Liquid/Solid Systems Comments/refs.)
24711   LSF (Linear Search Formula)
    0   LSFD (Liquid/Solid Systems Further Details)
 1073   LSPA (Liquid/Solid Systems Partner)
    8   LSPRES (Liquid/Solid Systems Pressure)
   14   LSSOLV (Liquid/Solid Systems Solvent)
  357   LSTEMP (Liquid/Solid Systems Temperature)
 1359   LV (Liquid/Vapor Systems)
 1359   LVC (Liquid/Vapor Systems Comments/refs.)
    0   LVFD (Liquid/Vapor Systems Further Details)
  328   LVPA (Liquid/Vapor Systems Partner)
  172   LVPRES (Liquid/Vapor Systems Pressure)
   76   LVSOLV (Liquid/Vapor Systems Solvent)
  243   LVTEMP (Liquid/Vapor Systems Temperature)
 1432   MCBS (Boundary Surface Phenomena)
 1432   MCBSC (Boundary Surface Phenomena Comments/refs.)
    0   MCBSFD (Boundary Surface Phenomena Further Details)
  439   MCBSPA (Boundary Surface Phenomena Partner)
    6   MCBSPRES (Boundary Surface Phenomena Pressure)
  319   MCBSSOLV (Boundary Surface Phenomena Solvent)
  354   MCBSTEMP (Boundary Surface Phenomena Temperature)
    0   MCE
 1811   MCEC (Energy Data Comments/refs.)
 1811   MCEDATA (Energy Data)
    0   MCEFD (Energy Data Further Details)
  207   MCEPART (Energy Data Partner)
   12   MCEPRES (Energy Data Pressure)
  142   MCESOLV (Energy Data Solvent)
  173   MCETEMP (Energy Data Temperature)
    1   MCMF (Multicomponent System Molecular Formula)
  879   MCOM (Other Mechanical Properties)
```

Appendix D   Tabulation of Occurrences of Data Fields

879	MCOMC (Other Mechanical Properties Comments/refs.)
0	MCOMFD (Other Mechanical Properties Further Details)
134	MCOMPART (Other Mechanical Properties Partner)
10	MCOMPRES (Other Mechanical Properties Pressure)
35	MCOMSOLV (Other Mechanical Properties Solvent)
117	MCOMTEMP (Other Mechanical Properties Temperature)
0	MCSYSTEM (Multicomponent System)
1338	MCTP (Transport Phenomena)
1338	MCTPC (Transport Phenomena Comments/refs.)
0	MCTPFD (Transport Phenomena Further Details)
381	MCTPPA (Transport Phenomena Partner)
26	MCTPPRES (Transport Phenomena Pressure)
374	MCTPSOLV (Transport Phenomena Solvent)
484	MCTPTEMP (Transport Phenomena Temperature)
0	MI (Moment of Inertia)
237	MIC (Moment of Inertia Comments/refs.)
0	MOLPOL (Molar Polarization)
0	MOLPOLC (Molar Polarization Comments/refs.)
1832988	MP (Melting Point)
1873718	MPC (Melting Point Comments/refs.)
498	MPOLC
634733	MPSOLV (Melting Point Solvent)
0	MRC (Rotational Constants)
629	MRCC (Rotational Constants Comments/refs.)
30146	MSC (Mass Spectrum Fragmentation Comments/refs.)
47	MSFRAG (Mass Spectrum Fragmentation)
416	MSMETH (Mass Spectrum Method)
0	MSMOL (Mass Spectrum Molecular Peak)
467	MSUS (Magnetic Susceptibility)
1872	MSUSC (Magnetic Susceptibility Comments/refs.)
39	MSUSTEMP (Magnetic Susceptibility Temperature)
1053	MUT (Mutarotation)
1465	MUTC (Mutarotation Comments/refs)
820	MUTCONC (Mutarotation Concentration (c,p))
6	MUTLEN (Mutarotation Length of)
1064	MUTSOLV (Mutarotation Solvent)
1070	MUTTEMP (Mutarotation Temperature)
867	MUTTIME (Mutarotation Time)
1089	MUTTYPE (Mutarotation Type)
1055	MUTWL (Mutarotation Wavelength)
12	MVOL (Molar Volume)
759	MVOLC (Molar Volume Comments/refs.)
4	MVOLPRES (Molar Volume Pressure)
30	MVOLTEMP (Molar Volume Temperature)

Appendix D   Tabulation of Occurrences of Data Fields

17719	NMRAC (Nuclear Magnetic Resonance Abs. Comments/refs.)
1586	NMRANUC (Nuclear Magnetic Resonance Abs. Nucleus)
815	NMRASOLV (Nuclear Magnetic Resonance Abs. Solvent)
12904	NMRSC (Nuclear Magnetic Resonance Sp. Comments/refs.)
567	NMRSNUC (Nuclear Magnetic Resonance Sp. Nucleus)
314	NMRSSOLV (Nuclear Magnetic Resonance Sp. Solvent)
0	NQC (Nuclear Quadrupole Coupling Constants)
475	NQCC (Nuclear Quadrupole Coupling Con. Comments/refs.)
103	NQCNUC (Nuclear Quadrupole Coupling Constants Nuclei)
0	OA (Optical Anisotropy)
437	OAC (Optical Anisotropy Comments/refs.)
500	ORD (Optical Rotary Dispersion)
4957	ORDC (Optical Rotary Dispersion Comments/refs.)
533	ORDSOLV (Optical Rotary Dispersion Solvent)
211658	ORP (Optical Rotatory Power)
212921	ORPC (Optical Rotatory Power Comments/refs.)
80820	ORPCONC (Optical Rotatory Power Concentration)
350	ORPLEN (Optical Rotatory Power Length of)
105163	ORPSOLV (Optical Rotatory Power Solvent)
211670	ORPTEMP (Optical Rotatory Power Temperature)
212511	ORPTYPE (Optical Rotatory Power Type)
211684	ORPW (Optical Rotatory Power Wavelength)
6	PHM (Phosphorescence Maxima Wavelength)
43	PHMC (Phosphorescence Maxima Comments/refs.)
9	PHMSOLV (Phosphorescence Maxima Solvent)
12	PHS (Phosphorescence Spectrum Wavelength)
39	PHSC (Phosphorescence Spectrum Comments/refs.)
0	PHSMETH (Phosphorescence Spectrum Excitation Method)
30	PHSSOLV (Phosphorescence Spectrum Solvent)
0	PHWP (Polarographic Half-wave Potential)
6239	PHWPC (Polarographic Half-wave Potential Comments/refs.)
0	PHWPSOLV (Polarographic Half-wave Solvent)
29186	PRBYPR (Preparation By-products)
2619029	PRC (Preparation Comments/refs.)
0	PRCAT (Preparation Catalyst)
119858	PRCOND (Preparation Other Conditions)
0	PRIRRAD (Preparation Irradiation)
6614	PRPRES (Preparation Pressure)
0	PRREFLUX (Preparation Reflux)
316634	PRRGT (Preparation Reagents)
0	PRRT (Preparation Ambient Temperature)
0	PRSOLV (Preparation Solvent)
394941	PRSTART (Preparation Starting Material)
61959	PRTEMP (Preparation Temperature)

Appendix D Tabulation of Occurrences of Data Fields

0	PRTIME (Preparation Time)
0	PRYIELD (Preparation Yield)
0	PUR (Purification (Method))
17979	PURC (Purification (Method) Comments/refs.)
0	PURS (Purity)
7189	PURSC (Purity Comments/refs.)
0	QM (Quadrupole Moment)
65	QMC (Quadrupole Moment Comments/refs.)
561	RAMANB (Raman Bands)
1387	RAMANBC (Raman Bands Comments/refs.)
561	RAMANS (Raman Spectrum)
4093	RAMANSC (Raman Spectrum Comments/refs.)
0	REDOX (Redox Potential)
1369	REDOXC (Redox Potential Comments/refs.)
0	REDOXSOL (Redox Potential Solvent)
262967	RI (Refractive Index)
263523	RIC (Refractive Index Comments/refs.)
262980	RITEMP (Refractive Index Temperature)
262980	RIW (Refractive Index Wavelength)
795	RTAFF (Affinity)
795	RTAFFC (Affinity Comments/refs.)
17471	RTANAL (Elementary Analysis)
17471	RTANALC (Elementary Analysis Comments/refs.)
5688	RTCAL (Calorific Data)
5688	RTCALC (Calorific Data Comments/refs.)
432	RTCOMP (Compressibility)
432	RTCOMPC (Compressibility Comments/refs.)
22636	RTCONF (Conformation)
22636	RTCONFC (Conformation Comments/refs.)
10662	RTCP (Crystal Phase)
10662	RTCPC (Crystal Phase Comments/refs.)
10789	RTCPL (Coupling Constants)
10789	RTCPLC (Coupling Constants Comments/refs.)
3516	RTEL (Electric Data)
3516	RTELC (Electric Data Comments/refs.)
7525	RTELCH (Electrochemical Behavior)
7525	RTELCHC (Electrochemical Behavior Comments/refs.)
469	RTELPOL (Electrical Polarizability)
469	RTELPOLC (Electrical Polarizability Comments/refs.)
1441	RTELS (Electronic Spectrum)
1441	RTELSC (Electronic Spectrum Comments/refs.)
1226	RTEMOL (Molecular Energy)
1226	RTEMOLC (Molecular Energy Comments/refs.)
4127	RTEMS (Emission Spectrum)

4127	RTEMSC (Emission Spectrum Comments/refs.)
3885	RTESR (ESR Data)
3885	RTESRC (ESR Data Comments/refs.)
103	RTGAS (Association in the Gas Phase)
103	RTGASC (Association in the Gas Phase Comments/refs.)
1314	RTLIQ (Liquid Phase)
1314	RTLIQC (Liquid Phase Comments/refs.)
512	RTMAG (Magnetic Data)
512	RTMAGC (Magnetic Data Comments/refs.)
3739	RTMDEF (Molecular Deformation)
3739	RTMDEFC (Molecular Deformation Comments/refs.)
800	RTMEC (Mechanical Properties)
800	RTMECC (Mechanical Properties Comments/refs.)
4088	RTNMR (NMR Data)
4088	RTNMRC (NMR Data Comments/refs.)
2683	RTNQR (Nuclear Quadrupole Resonance)
2683	RTNQRC (Nuclear Quadrupole Resonance Comments/refs.)
4657	RTOPT (Optics)
4657	RTOPTC (Optics Comments/refs.)
5366	RTOSM (Other Spectroscopic Methods)
5366	RTOSMC (Other Spectroscopic Methods Comments/refs.)
2343	RTROTS (Rotational Spectrum)
2343	RTROTSC (Rotational Spectrum Comments/refs.)
1149449	RTSF (Short File Keyword)
1149449	RTSFC (Short File Keyword Comments/refs.)
6651	RTSKEL (Interatomic Distances and Angles)
6651	RTSKELC (Interatomic Distances and Angles Comments/refs)
555	RTUP (Ultrasonic Properties)
555	RTUPC (Ultrasonic Properties Comments/refs.)
2250	RTVIBS (Vibrational Spectrum)
2250	RTVIBSC (Vibrational Spectrum Comments/refs.)
77566	RXNAIM (Chemical Reaction Aim of the Study)
377609	RXNC (Chemical Reaction Comments/refs.)
0	RXNCAT (Chemical Reaction Catalyst)
26011	RXNCOND (Chemical Reaction Further Conditions)
0	RXNIRRAD (Chemical Reaction Irradiation)
58334	RXNPART (Chemical Reaction Reaction Partner)
1908	RXNPRES (Chemical Reaction Pressure)
57782	RXNPROD (Chemical Reaction Reaction Product)
0	RXNREFLX (Chemical Reaction Reflux)
0	RXNRGT (Chemical Reaction Reagents)
0	RXNRT (Chemical Reaction Ambient Temperature)
0	RXNSOLV (Chemical Reaction Solvent)
21414	RXNTEMP (Chemical Reaction Temperature)

Appendix D Tabulation of Occurrences of Data Fields

0	RXNTIME (Chemical Reaction Time)
0	SDELTA (Entropy Delta S)
2370	SDELTAC (Entropy Delta S Comments/refs.)
0	SDELTAP (Entropy Delta S Pressure)
0	SDELTAT (Entropy Delta S Temperature)
0	SDIF (Self-diffusion)
261	SDIFC (Self-diffusion Comments/refs.)
38	SDIFTEMP (Self-diffusion Temperature)
11	SFOR (Entropy of Formation)
172	SFORC (Entropy of Formation Comments/refs.)
0	SFORPRES (Entropy of Formation Pressure)
5	SFORTEMP (Entropy of Formation Temperature)
2009	SL (Solubility)
23274	SLC (Solubility Comments/refs.)
1	SLP (Solubility Product)
88	SLPC (Solubility Product Comments/refs.)
0	SLPRATIO (Solubility Product Ratio of Solvents)
1	SLPSOLV (Solubility Product Solvent)
1	SLPTEMP (Solubility Product Temperature)
26	SLRATIO (Solubility Ratio of Solvents)
4377	SLSOLN (Solubility in Solution or in Pure Solvent)
4824	SLSOLV (Solubility Solvent)
3472	SLTEMP (Solubility Temperature)
2058	SOLNB (Solution Behavior)
2058	SOLNBC (Solution Behavior Comments/refs.)
0	SOLNBFD (Solution Behavior Further Details)
1025	SOLNBPA (Solution Behavior Partner)
131	SOLNBPR (Solution Behavior Pressure)
225	SOLNBSOL (Solution Behavior Solvent)
543	SOLNBT (Solution Behavior Temperature)
2185	ST (Surface Tension)
3283	STC (Surface Tension Comments/refs.)
0	STRUCT (Related Structure)
144154	STRUCTC (Related Structure Comments/refs.)
2186	STTEMP (Surface Tension Temperature)
2805	SUB (Sublimation)
10767	SUBC (Sublimation Comments)
2418	SUBPRES (Sublimation Pressure)
1375762	SY (Synonym or Trivial Name)
0	TCON (Thermal Conductivity)
411	TCONC (Thermal Conductivity Comments/refs.)
224	TCONTEMP (Thermal Conductivity Temperature)
74	TP (Triple Point)
150	TPC (Triple Point Comments/refs.)

Appendix D   Tabulation of Occurrences of Data Fields

  1360 TX (Toxicity)
  1360 TXC (Toxicity Comments/refs.)
  1685 US (Use)
  1685 USC (Use Comments/refs.)
    19 VISB (Bulk Viscosity)
    36 VISBC (Bulk Viscosity Comments/refs.)
    29 VISBTEMP (Bulk Viscosity Temperature)
  1815 VISD (Dynamic Viscosity)
  2000 VISDC (Dynamic Viscosity Comments/refs.)
  1895 VISDTEMP (Dynamic Viscosity Temperature)
   601 VISK (Kinematic Viscosity)
   913 VISKC (Kinematic Viscosity Comments/refs.)
   645 VISKTEMP (Kinematic Viscosity Temperature)
  1190 VP (Vapor Pressure)
  3225 VPC (Vapor Pressure Comments/refs.)
  1420 VPTEMP (Vapor Pressure Temperature)

References

1 *Searching DIALOG. The Complete Guide*, Dialog Information Services Inc., Palo Alto, California, 1987.

Index

P

parameters of crystal lattices, III-82
parent structures in ROSDAL, V-12
parentheses, use of, II-12
partial chemical names, III-5, III-9
partial molecular formula, III-25
PARTNER field, III-44, III-56
partner in adsorption, III-130
partner in association, III-131
patent numbers, III-64
patents, III-62
PC field, III-79
PCPLOT, V-22
PCs, use of, V-22
periodic table, rows and columns in,
   III-27
periodic table, searches in, III-27
phase transition, enthalpy of, III-181,
   III-187
PHM field, III-212
phosphorescence spectra, III-204, III-211
phosphorescence spectrum maximum,
   III-212
PHS field, III-211
PHWP field, III-169
physical properties data, II-14, IV-5, IV-6
physical state, III-66, IV-5
physiological data, III-132, III-133
pK values, III-162
PN field, III-62, III-64
polarizability, electrical, III-84
polarizability, molecular, III-84, III-141
polarized light, III-16
polarographic half-wave potential, III-162,
   III-169
potential, ionization, III-84, III-107
potential, molecular, IV-15
potential, polarographic half-wave, III-169
potential, redox, III-162, III-169
potentiometry in dissociation constants,
   III-165
power of optical rotation, III-148
power, optical rotatory, III-150

PR format, III-44, IV-9
Prager, B., I-2
preparative data, II-14, III-5
precursors, chemical, III-44
predefined formats, II-16
preparations data, output of, IV-9
preparative data, chemical names in, III-7,
   III-44
presence of elements, III-24
pressure at sublimation temperature,
   III-76
pressure in enthalpy of vaporization,
   III-186
pressure of boiling point, III-71
pressure of reactions, III-44
pressure-surface isotherms, III-129
pressure, critical, III-66
pressure, heat capacity at constant,
   III-181
PRINT command, II-14
printed *Handbook*, III-58
printers, use of with GEOFF, V-30
printing answers, II-13
printing structures, V-30
printing structures with GEOFF, V-30
printing, offline, II-14
priority among operators, II-12
PRODUCT field, III-44, III-56
product of hydrogenation, III-181
product, solubility, III-110
products, III-6
prompt, system, II-3, III-2
properties data output, IV-5
properties, critical, III-66, III-68
properties, crystal, III-66, III-79
proteins, III-167
proton affinity, III-84, III-108
proximity operator, III-15, III-71, III-89,
   III-107, III-111, III-122, III-142, III-149,
   III-168, III-180, III-187
proximity operators, II-10, III-2
proximity searching, III-7
PS format, III-16
PUR field, IV-51